ISBN 978-1-332-17884-1
PIBN 10294419

PLANNING AND CODING OF PROBLEMS
FOR AN
ELECTRONIC COMPUTING INSTRUMENT

BY

Herman H. Goldstine   John von Neumann

Report on the Mathematical and Logical aspects of an
Electronic Computing Instrument

Part II, Volume I - 3

Institute for Advanced Study
Princeton, New Jersey
1947

TABLE OF CONTENTS

PREFACE

This report was prepared in accordance with the terms of Contract No. W-36-034-ORD-7481 between the Research and Development Service, U. S. Army Ordnance Department and the Institute for Advanced Study. It is essentially the second paper referred to in the Preface of the earlier report entitled, "Preliminary Discussion of the Logical Design of an Electronic Computing Instrument", by A. W. Burks, H. H. Goldstine and John von Neumann, dated 28 June 1946.

During the time which has intervened it has become clear that the issuance of a greater number of reports will be desirable. In this sense the report, "Preliminary Discussion of the Logical Design of an Electronic Computing Instrument" should be viewed as Part I of the sequence of our reports on the Electronic Computer. The present report is Volume I of Part II. A Volume II of Part II will follow within a few months.

Part II is intended to give an account of our methods of coding and of the philosophy which governs it. It contains what we believe to be an adequate number of typical examples, going from the simplest case up to several of high complexity, and in reasonable didactic completeness. On the other hand, Part II in its present form is preliminary in certain significant ways. On this account the following ought to be said.

We do not discuss the orders controlling the inputs and outputs and the stopping of the machine. (The inputs and outputs are, however, discussed to the extent which is required for the treatment of the binary-decimal and decimal-binary conversions in Chapter VIII.) The reason for this is that in this direction some engineering alternatives seem worth keeping open, and it does not affect the major problems of coding relevantly.

The use of the input-output medium as a subsidiary memory is not discussed. This will probably be taken up in Part III.

The code is not final in every respect. Several minor changes are clearly called for, and even some major ones are conceivable, depending upon various engineering possibilities that are still open. In all cases where changes are indicated or possible, there are several alternative ones, between which it is not yet easy to choose with much assurance. We believe that our present code is correct in its basic idea, and at any rate a reasonable basis for the discussion of other proposals. That is, the examples which we have coded, may serve as standards in comparing this code with variants or with more basically different systems.

It should be noted that in comparing codes, four viewpoints must be kept in mind, all of them of comparable importance: (1) Simplicity and reliability of the engineering solutions required by the code; (2) Simplicity, compactness and completeness of the code; (3) Ease and speed of the human procedure of translating mathematically conceived methods into the code, and also of finding and correcting errors in coding or of applying to it changes that have been decided upon at a late stage; (4) Efficiency of the code in operating the machine near its full intrinsic speed.

We propose to carry our comparisons of various variants and codes under these aspects.

The authors wish to express their thanks to Professor Arthur W. Burks, of the University of Michigan, for many valuable discussions, as well as for his help in coding certain of the problems in the text.

H. H. Goldstine
J. von Neumann

Institute for Advanced Study
1 April 1947

PLANNING AND CODING OF PROBLEMS FOR AN ELECTRONIC COMPUTING INSTRUMENT

7.0 GENERAL PRINCIPLES OF CODING AND FLOW-DIAGRAMMING.

7.1 In the first part of this report we discussed in broad outline our basic point of view in regard to the electronic computing machine we now envision. There is included in that discussion a tabulation of the orders the machine will be able to obey. In this, the second part of the report, we intend to show how the orders may be used in the actual programming of numerical problems.

Before proceeding to the actual programming of such problems, we consider it desirable to discuss the nature of coding per se and in doing this to lay down a modus operandi for handling specific problems. We attempt therefore in this chapter to analyze the coding of a problem in a detailed fashion, to show where the difficulties lie, and how they are best resolved.

The actual code for a problem is that sequence of coded symbols (expressing a sequence of words, or rather of half-words and words) that has to be placed into the selectron memory in order to cause the machine to perform the desired and planned sequence of operations, which amount to solving the problem in question. Or to be more precise: This sequence of codes will impose the desired sequence of actions on C by the following mechanism: C scans the sequence of codes, and effects the instructions, which they contain, one by one. If this were just a linear scanning of the coded sequence, the latter remaining throughout the procedure unchanged in form, then matters would be quite simple. Coding a problem for the machine would merely be what its name indicates: Translating a meaningful text (the instructions that govern solving the problem under consideration) from one language (the language of mathematics, in which the planner will have conceived the problem, or rather the numerical procedure by which he has decided to solve the problem) into another language (that one of our code).

This, however, is not the case. We are convinced, both on general grounds and from our actual experience with the coding of specific numerical problems, that the main difficulty lies just at this point. Let us therefore describe the process that takes place more fully.

The control scans the coded instructions in the selectron memory as a rule linearly, i.e. beginning, say, with word No. 0, and proceeding from word No. y to word No. y + 1 (and within each word from its first half to its second half), but there are exceptions: The transfer orders xC, xC', xCc, xCc' (Cf. Table I at the end of the first part of the report, or Table II below) cause C to jump from word y to the arbitrarily prescribed word x (unconditionally or subject to the fulfillment of certain conditions). Also, these transfer orders are among the most critical constituents of a coded sequence. Furthermore, the substitution orders xSp, xSp' (and also xS, cf. as above) permit C to modify any part of the coded sequence as it goes along. Again, these substitution orders are usually of great importance.

To sum up: C will, in general, not scan the coded sequence of instructions linearly. It may jump occasionally forward or backward, omitting (for the time being, but probably not permanently) some parts of the sequence, and going repeatedly through others. It may modify some parts of the sequence while obeying the instructions in another part of the sequence. Thus when it scans a part of the sequence several times, it may actually find a different set of instructions there at each passage. All these displacements and modifications may be conditional upon the nature of intermediate results obtained by the machine itself in the course of this procedure. Hence it will not be possible in general to foresee in advance and completely the actual course of C, its character and the sequence of its omissions on one hand and of its multiple passages over the same place on the other, as well as the actual instructions it finds along this course, and their changes through various successive occasions at the same place, if that place is multiply traversed by the course of C. These circumstances develop in their actually assumed forms only during the process (the calculation) itself, i.e. while C actually runs through its gradually unfolding course.

Thus the relation of the coded instruction sequence to the mathematically conceived procedure of (numerical) solution is not a statical one, that of a translation, but highly dynamical: A coded order stands not simply for its present contents at its present location, but more fully for any succession of passages of C through it, in connection with any succession of modified contents to be found by C there, all of this being determined by all other orders of the sequence (in conjunction with the one now under consideration) This entire, potentially very involved, interplay of interactions evolves successively while C runs through the operations controlled and directed by these continuously changing instructions.

These complications are, furthermore, not hypothetical or exceptional. It does not require a deep analysis of any inductive or iterative mathematical process to see that they are indeed the norm. Also, the flexibility and efficiency of our code is essentially due to them, i.e. to the extensive combinatorial possibilities which they indicate. Finally, those mathematical problems, which by their complexity Justify the use of the machine that we envision, require an application of these control procedures at a rather high level of complication and of multiplicity of the course of C and of the successive changes in the orders.

All these assertions will be amply justified and elaborated in detail by the specific coded examples which constitute the bulk of this report, and by the methods that we are going to evolve to code them.

Our problem is then to find simple, step-by-step methods, by which these difficulties can be overcome. Since coding is not a static process of translation, but rather the technique of providing a dynamic background to control the automatic evolution of a meaning, it has to be viewed as a logical problem and one that represents a new branch of formal logics. We propose to show in the course of this report how this task is mastered.

The balance of this chapter gives a rigorous and complete description of our method of coding, and of the auxiliary concepts which we found convenient to introduce in order to expound this method. (The subsequent chapters of the report deal with specific examples, and with the methods of combining already existing coded sequences.) Since this is the first report on this subject, we felt Justified to stress rigor and completeness rather than didactic simplicity. A later presentation will be written from the didactic point of view, which, we believe, will show that our methods are fairly easy to learn.

7.2 Table II on page 5 is essentially a repetition of Table I at the end of Part I of this report, i.e. it is a table of orders with their abbreviations and explanations. We found it convenient to make certain minor changes. They are as follows:

First: 11 has been changed, so as to express the round off rule discussed in 5.11 in Part I of this report. In accord with the discussion carried out loc. cit., we use the first round off rule described there. As far as the left-hand 39 digits are concerned, this consists of adding one to digit 40, and effecting the resulting carries, thereby possibly modifying the left-hand 39 digits. Since we keep track of the right-hand 39 digits, too, it is desirable to compensate for this within the extreme left (non sign) digit of the right-hand 39 digits. This amounts to adding $\frac{1}{2}$ in the register. The number there is $\geqq 0$ (sign digit 0), hence a carry results if and only if that number is $\geqq \frac{1}{2}$, i.e. if its extreme left (non sign) digit is 1. In this case the carry adds 2^{-39} in the accumulator and subtracts 1 in the register. Hence instead of adding $\frac{1}{2}$ in the register we actually subtract $\frac{1}{2}$ there. However, the aggregate result (on the 78 digit number) is in any event the addition of $\frac{1}{2}$ in the register (i.e. of one to digit 40), hence it must be compensated by subtracting $\frac{1}{2}$ in the register. Consequently we do nothing or subtract 1 in the register, according to whether its extreme left (non sign) digit is 0 or 1. And, as we saw above, we correspondingly do nothing or add 2^{-39} in the accumulator Note, that the operation $+2^{-39}$, which may cause carries, takes place in the accumulator, which can indeed effect carries; while the operation -1, which can cause no carries, takes place in the register, which cannot effect carries. Indeed: Subtracting 1 in the register, where the sign digit is 0 (cf. above) merely amounts to replacing the sign digit by 1. The new formulation of 11 expresses exactly these rules.

Second: 18, 19 have been changed somewhat for reasons set forth in 8.2 below. We note that 18, 19 assume, both in their old form (Table I) and their new form (Table II), that in each order the memory position number x occupies the 12 left digits (cf. 6.6.5 in Part I of this report).

Third: 20, 21 have also been changed. The new form of 20 is a right shift, which is so arranged that it halves the number in the accumulator, including an arithmetically correct treatment of the sign digit. The new form of 21 is a left shift, which is so arranged that it doubles the number in the accumulator (provided that that number lies between $-\frac{1}{2}$ and $\frac{1}{2}$), including an arithmetically correct treatment of the sign digit. At the same time, however, 21 (in its new form) is so arranged that the extreme left digit (after the sign digit) in the accumulator is not lost, but transferred into the register. It is inserted there at the extreme right, and the original contents of the register are shifted left correspondingly. The immediate uses of 20, 21 are arithmetical, but the last mentioned features of 21 are required for other, rather combinatorial uses of 20, 21. For these uses there will be some typical examples in 9.5 and 9.6.

It should be added, that various elaborations and refinements of 20, 21 might be considered. Shifts by a given number of places, shifts until certain sizes have been reached, etc. We do not consider the time mature as yet as to make any definite choices in this respect, although some variants would have certain advantages in various situations. We will, for the time being, use the simplest forms of 20, 21 as given in Table II.

7.3 We now proceed to analyze the procedures by which one can build up the appropriate coded sequence for a given problem — or rather for a given numerical method to solve that problem. As was pointed out in 7.1, this is not a mere question of translation (of a mathematical text into a code), but rather a question of providing a control scheme for a highly dynamical process, all parts of which may undergo repeated and relevant changes in the course of this process.

Let us therefore approach the problem in this sense.

It should be clear from the preliminary analysis of 7.1, that in planning a coded sequence the thing that one should keep primarily in mind is not the original (initial) appearance of that sequence, but rather its functioning and continued changing while the process that it controls goes through its course. It is therefore advisable to begin the planning at this end, i.e. to plan first the course of the process and the relationship of its successive stages to their changing codes, and to extract from this the original coded sequence as a secondary operation. Furthermore, it seems equally clear, that the basic feature of the functioning of the code in conjunction with the evolution of the process that it controls, is to be seen in the course which the control C takes through the coded sequence, paralleling the evolution of that process. We therefore propose to begin the planning of a coded sequence by laying out a schematic of the course of C through that sequence, i.e. through the required region of the selectron memory. This schematic is the *flow diagram of C*. Apart from the above a priori reasons, the decision to make the flow diagram of C the first step in code-planning appears to be extensively justified by our own experience with the coding of actual problems. The exemplification of this kind of experience on a number of selected typical examples forms the bulk of this report.

In drawing the flow diagram of C the following points are relevant:

First: The reason why C may have to move several times through the same region in the selectron memory is that the operations that have to be performed may be repetitive. Definitions by induction (over an integer variable); iterative processes (like successive approximations); calculations of sums or of products of many addends or factors all formed according to the same law (but depending on a variable summation or multiplication index); stepwise integrations in a simple quadrature or a more involved differential equation or system of differential equations, which approximate the result and which are iterative in the above sense; etc. — these are typical examples of the situation that we have in mind. To simplify the nomenclature, we will call any simple iterative process of this type an *induction* or a *simple induction*. A multiplicity of such iterative processes, superposed upon each other or crossing each other will be called a *multiple induction*.

When a simple induction takes place, C travels during each step of the induction over a certain path, at the end of which it returns to its beginning. Hence this path may be visualized as a loop. We will call it an *induction loop* or a *simple induction loop*. A multiple induction gives rise to a multiplicity of such loops; they form together a pattern which will be called a ***multiple induction loop***.

We propose to indicate these portions of the flow diagram of C by a symbolism of lines oriented by arrows. Thus a linear sequence of operations, — with no inductive elements in it, will be denoted by symbols like those in Figures 7.1 a-c, while a simple induction loop is shown in d, eod., and multiple induction loops are shown in e-f, eod.

	Complete	Abbreviation	
1.	$S(x) \rightarrow Ac$	x	Clear accumulator and add number located at position x in the selectrons into it.
2.	$S(x) \rightarrow Ac-$	x –	Clear accumulator and subtract number located at position x in the selectrons into it.
3.	$S(x) \rightarrow AcM$	x M	Clear accumulator and add absolute value of number located at position x in the selectrons into it.
4.	$S(x) \rightarrow Ac-M$	x –M	Clear accumulator and subtract absolute value of number located at position x in the selectrons into i
5.	$S(x) \rightarrow Ah$	x h	Add number located at position x in the selectrons into the accumulator.
6.	$S(x) \rightarrow Ah-$	x h–	Subtract number located at position x in the selectrons into the accumulator.
7.	$S(x) \rightarrow AhM$	x hM	Add absolute value of number located at position x in the selectrons into the accumulator.
8.	$S(x) \rightarrow Ah-M$	x h–M	Subtract absolute value of number located at position x in the selectrons into the accumulator.
9.	$S(x) \rightarrow R$	x R	Clear register* and add number located at position x in the selectrons into it.
10.	$R \longrightarrow A$	A	Clear accumulator and shift number held in register into it.
11.	$S(x)$ $XR \rightarrow A$	x X	Clear accumulator and multiply the number located at position x in the selectrons by the number in the register, placing the left-hand 39 digits of the answer in the accumulator and the right-hand 39 digits of the answer in the register. The sign digit of the register is to be made equal to the extreme left (non sign) digit. If the latter is 1, then 2^{-39} is to be added into the accumulator.
12.	$A \div S(x) \rightarrow R$	x ÷	Clear register and divide the number in the accumulator by the number located in position x of the selectrons, leaving the remainder in the accumulator and placing the quotient in the register.
13.	$Cu \rightarrow S(x)$	x C	Shift the control to the left-hand order of the order pair located at position x in the selectrons.
14.	$Cu' \rightarrow S(x)$	x C'	Shift the control to the right-hand order of the order pair located at position x in the selectrons.
15.	$Cc \rightarrow S(x)$	x Cc	If the number in the accumulator is ≥ 0, shift the control as in $Cu \rightarrow S(x)$.
16.	$Cc' \rightarrow S(x)$	x Cc'	If the number in the accumulator is ≥ 0, shift the control as in $Cu' \rightarrow S(x)$.
17.	$At \rightarrow S(x)$	x S	Transfer the number in the accumulator to position x in the selectrons.
18.	$Ap \rightarrow S(x)$	x Sp	Replace the left-hand 12 digits of the left-hand order located at position x by the 12 digits 9 to 20 (from the left) in the accumulator.
19.	$Ap' \rightarrow S(x)$	x Sp'	Replace the left-hand 12 digits of the right-hand order located at position x by the 12 digits 29 to 40 (from the left) in the accumulator.
20.	R	R	Replace the contents $\xi_0\ \xi_1\ \xi_2 \dots \xi_{38}\ \xi_{39}$ of the accumulator by $\xi_0\ \xi_0\ \xi_1 \dots \xi_{37}\ \xi_{38}$.
21.	L	L	Replace the contents $\xi_0\ \xi_1\ \xi_2 \dots \xi_{38}\ \xi_{39}$ and $\eta_0\ \eta_1\ \eta_2 \dots \eta_{38}\ \eta_{39}$ of the accumulator and the register by $\xi_0\ \xi_2\ \xi_3 \dots \xi_{39} 0$ and $\eta_1\ \eta_2\ \eta_3 \dots \eta_{39}\ \xi_1$.

*Register means arithmetic register.

FIGURE 7.1

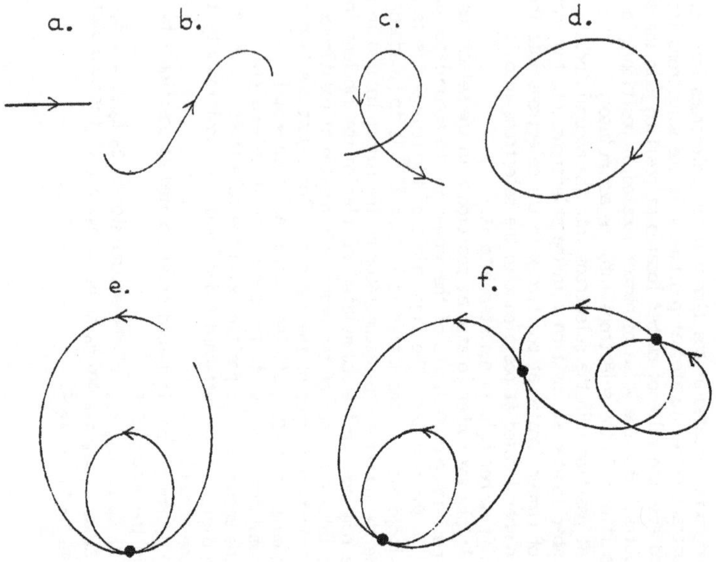

Second: It is clear that this notation is incomplete and unsatisfactory. Figures 7.1 d-f fail to indicate how C gets into these loops, how many times it circles each loop, and how it leaves it. e-f, eod. also leave it open what the hierarchy of the loops is, in what order they are taken up by C, etc.

Actually the description of an induction is only complete if a criterium is specified which indicates whether the iterative step should be repeated, i.e. the loop circled once more, or whether the iterations are completed and a new path is to be entered. Accordingly a simple induction loop, like the one shown in Figure 7.1 d, needs an indication of the path on which C enters it at the beginning of the induction, the path on which C leaves it at the end of the induction, and the area in which the criterium referred to above is applied, i.e. where it is determined whether C is to circle the loop again, or whether it is to proceed on the exit path. We will denote this area by a box with one input arrow and two output arrows, and we call this an *alternative box*. Thus Figure 7.1 d, becomes the more complete Figure 7.2 b. The alternative box may also be used to bifurcate a linear, non-looped piece of C's course. Indeed, alternative procedures may be required in non-inductive procedures, too. This is shown in Figure 7.2 a. Finally multiple induction loops, completed in this sense, are shown on c-d, eod. It will be noted that in these inductions in c the small loop represents an induction that is subordinate to that one on the big loop, i.e. part of its inductive step. Similarly in d the two small loops that are attached to the big loop are in the same subordinate relation to it, while the loop at the extreme right is again subordinate to the loop to which it is attached.

FIGURE 7.2

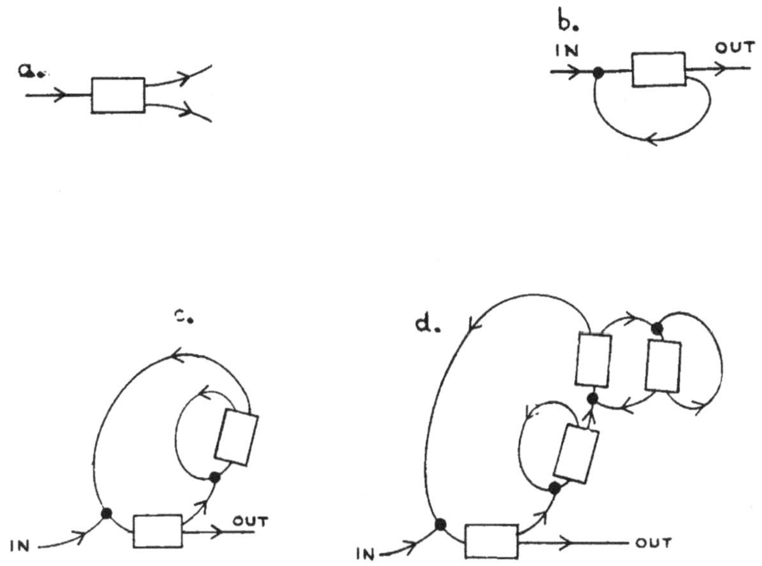

Third: The alternative boxes which we introduced correspond to the conditional transfer orders xCc, xCc'. I.e., the intention that they express will be effected in the actual code by such an order. Of the two output branches (cf. e.g. Figure 7.2 a) one leads to the order following immediately in the selectron memory upon the last order on the input branch, while the other leads to the left or the right hand order in S(x). If at the moment at which this decision is made the number u is in A, then u < o causes the first branch to be taken. We will place the u which is thus valid into the alternative box, and mark the two branches representing the two alternatives u ≧ o and u < o by + and by –, respectively. In this way Figures 7.2 a–b become Figures 7.3 a–b. Figure 7.3 b may be made still more specific: If the induction variable is i, and if the induction is to end when i reaches the value I (if i's successive values are 0, 1, 2 ..., then this means that I iterations are wanted), and if, as shown in Figure 7.3 b, the – branch is in the induction loop while the + branch leaves it, then the u of this Figure may be chosen as i – I, and the complete scheme is that shown in Figure 7.3 c. (In many inductions the natural ending is defined by i + 1, having reached a certain value I. Then the above i – I is to be replaced by j – I + 1)

FIGURE 7.3

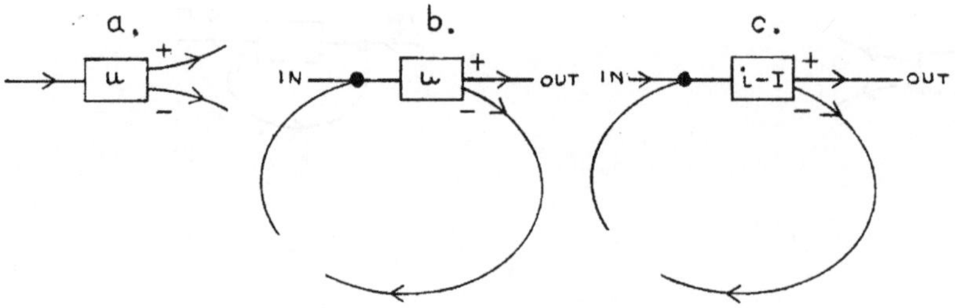

Fourth: Two or more paths of the flow diagram may merge at certain points. This takes place of necessity at the input of the alternative box of an induction loop (cf. Figures 7.2 b-d and 7.3 b-c), but it can also happen in a linear, non-looped piece of C's course, as shown in Figure 7.4. This corresponds to two alternative procedures leading up to a common continuation.

FIGURE 7.4

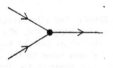

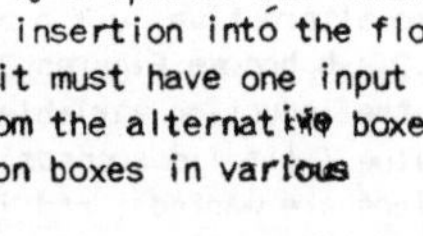

7 4 The flow diagram of C, as described in 7.3 is only a skeleton of the process that is to be represented. We pass therefore to examining the additions which have to be made in order to complete the scheme.

First: Our flow diagram in its present form does not indicate what arithmetical operations and transfers of numbers actually take place along the various parts of its course. These, however, represent the properly mathematical (as distinguished from the logical) activities of the machine, and they should be shown as such. For this reason we will denote each area in which a coherent group of such operations takes place by a special box which we call an *operation box*. Since a box of this category is an insertion into the flow diagram at a point where no branching or merger takes place, it must have one input arrow and one output arrow. This distinguishes it clearly from the alternative boxes of Figure 7 3. Figures 7.5 a-c show the positions of operation boxes in various looped and unlooped flow diagrams.

FIGURE 7.5

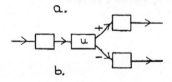

Second: While C moves along the flow diagram, the contents of the alternative boxes (which we have indicated) and of the operation boxes (which we have not indicated yet) will, in general, keep changing — and so will various other things that have to be associated with these in order to complete the picture (and which we have not yet indicated either). This whole modus procedendi assumes, however, that the flow diagram itself (i.e. the skeleton of Figure 7.3) remains unchanged. This, however, is not unqualifiedly true. To be specific: The transfer orders xC, xC' (unconditional), and also xCc, xCc' (conditional), can be changed in the course of the process by substitution orders xSp, xSp'. This changes the connections between various parts of the flow diagram. When this happens, we will terminate the corresponding part of the flow diagram with a circle marked with a Greek letter, and originate the alternative continuations from circles marked with properly indexed specimens of the same Greek letter, as shown in Figure 7.6 b. We call this arrangement a *variable remote connection.*

FIGURE 7.6

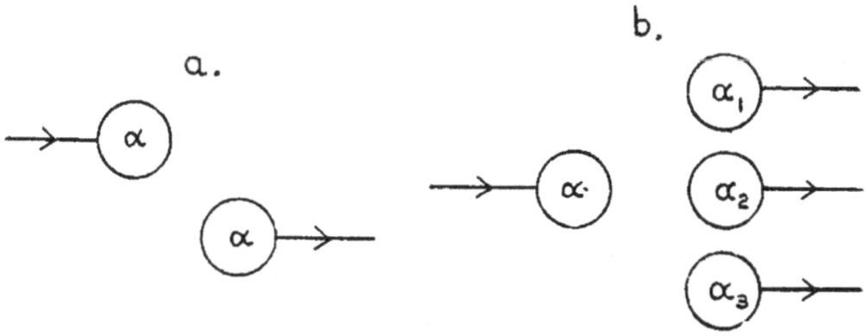

The circle  is the *entrance* to this remote connection, the circles  are the *exits* from it.

NOTE 1: The lettering with Greek letters α, β, γ, ... need neither be monotone nor uninterrupted, and decimal fractions following the Greek letters may be used for further subdivisions or in order to interpolate omissions, etc. All of this serves, of course, to facilitate modifications and corrections on a diagram which is in the process of being drawn up. The same principles will be applied subsequently to the enumerations of other entities: Constancy intervals (cf. the last part of 7.6, the basic Greek letters being replaced by Arabic numerals), storage positions (cf. the beginning of 7.7, the basic Greek letters being replaced by English capitals), operation boxes and alternative boxes (cf. the last part of 7.8, the basic Greek letters being replaced by Roman numerals).

NOTE 2: In the case of an unconditional transfer order which is never substituted, the flow diagram can be kept rigid, and no remote connection need be used. It is nevertheless sometimes convenient to use one even in such a case as shown in Figure 7.6 a. Indeed, this device may be used at any point of the flow diagram, even at a point where no transfer order is located and C scans linearly (i.e. from one half-word to the next one). This is so when the flow diagram threatens to assume an unwieldy form (e.g. grow off the paper) and it is desired to dissect it or to rearrange it. An arrangement of this type will be called a *fixed remote connection.*

The circles $\rightarrow\!\!\bigcirc$ and $\bigcirc\!\!\rightarrow$ are again the *entrance* and the *exit* of this connection. The Greek letter at the exit need, of course, not be indexed now. It is sometimes convenient to give a remote connection of either type more than one entrance. These are then equivalent to a pre-entrance confluence. They may be used when the geometry of the flow diagram makes the explicit drawing of the confluence inconvenient.

7.5 We pointed out at the beginning of the second remark in 7.4, that our present form of the flow diagram is still incomplete in two major respects: First, the contents of the operation boxes have not yet been indicated. Second, certain further features of the process that accompanies the course of C have not yet been brought out. Let us begin by analyzing the latter point.

No complicated calculation can be carried out without *storing* considerable numerical material while the calculation is in progress. This *storage* problem has been considered in considerable detail in Part I of this report (cf. in particular I.3, 2, 4 there). The storage may involve data which are needed for the entire duration of the procedure, and are therefore never changed by that procedure, as well as data which play a role only during part of the procedure (i.e. during one circling of an induction loop, or during a part of the course over a linear piece or an induction loop) and may therefore be changed (i.e. replaced by other data) at the end of their usefulness. These changes serve, of course, the very important purpose of making it possible to use the same storage (selectron memory) space successively for changing and different ends We will talk accordingly of *fixed* and of *variable storage*. These concepts are to be understood as being meaningful only relatively to the procedure which we are coding. Occasionally, it will also be found useful to use them relatively to certain parts of that procedure.

Before we develop the means to indicate the form and contents of the (fixed and variable) storage that is required, it is necessary to go into another matter.

A mathematical-logical procedure of any but the lowest degree of complexity cannot fail to require *variables* for its description. It is important to visualize that these variables are of two kinds, namely: First, a kind of variable for which the variable that occurs in an induction (or more precisely: with respect to which the induction takes place) is typical. Such a variable exists only within the problem. It assumes a sequence of different values in the course of the procedure that solves this problem, and these values are successively determined by that procedure as it develops. It is impossible to substitute a value for it and senseless to attribute a value to it "from the outside". Such a variable is called (with a term borrowed from formal logics) a *bound variable*. Second, there is another kind of variable for which the parameters of the problem are typical -- indeed it is essentially the same thing as a parameter. Such a variable has a fixed value throughout the procedure that solves the problem, i.e. a fixed value for the entire problem. If it is treated as a variable in the process of planning the coded sequence, then a value has to be substituted for it and attributed to it ("from the outside"), in order to produce a coded sequence that can be actually fed into the machine. Such a variable is called (again, borrowing a term from formal logics) a *free variable*.

In discussing the ways to indicate the form and contents of the storage that is required, there is no trouble in dealing with free variables; They are as good as constants, and the fact that they will acquire specific values only after the coding process is completed does not influence or complicate that process. Also, fixed storage offers no problems. The real crux of the matter is the variable storage and its changes. An item in the variable storage changes when it is explicitly replaced by a different one. (This is effected by the operation boxes, cf. 7.4 and the detailed discussion in 7.7, 7.8.) When the value of a bound variable changes (this is effected by the substitution boxes, cf. the last part of 7.6), this should also cause indirectly a change of those variable storage items in whose expression this variable occurs. However, we prefer to treat these variable value changes merely as changes in notation which do not entail any actual change in the relevant variable storage items. On the contrary, their function is to establish agreement between preceding and following expressions (occupying identical storage positions), which differ as expressions, but whose difference is removed by the substitution in question. (For details cf. the third rule relative to storage in 7.8.)

7.6 After these preparations we can proceed to an explicit discussion of storage and of bound variables.

In order to be able to give complete indications of the state of these entities, it is necessary to keep track of their changes. We will therefore mark along the flow diagram the points where changes of the content of any variable storage or of the value or domain of variability of any bound variable takes place. These points will be called the *transition points*. The transition points subdivide the flow diagram into connected pieces, along each of which all changing storages have constant contents and all bound variables have constant values. (The remote connections of the second remark in 7.4 rate in this respect as if they were unbroken paths. For a variable remote connection all possible branches are regarded in this way.) We call these *constancy intervals*. Clearly a constancy interval is bounded by the transition points on its endings.

NOTE: The constant contents and values attached to a constancy interval may, of course, vary from one passage of C over that interval to another.

Let us now consider the transition points somewhat more closely. We introduced so far three kinds of interruptions of the flow diagram: Alternative boxes, operation boxes, remote connections. Of these the first and the third effect no changes of the type under consideration, they influence only the course of C. The second, however performs arithmetical operations, therefore it may require storage, and hence effect changes in the variable storage. In fact, all variable storage originates in such operations, and all operations of this type terminate in consigning their results to storage, which storage, just by virtue of this origin, must be variable. Accordingly the operation boxes are the transition points inasmuch as the changing of variable storage is concerned. Thus we are left to take care of those transition points which change the values or delimit the domains of variability of bound variables. Changing the value of a bound variable involves arithmetical operations which must be indicated. An inductive variable i, which serves to enumerate the successive stages of an iteration, undergoes at each change an increase by one. We denote this substitution by $i + 1 \rightarrow i$. It is, however, neither necessary nor advisable to permit no other substitutions. Indeed, even an induction or iteration begins with a different substitution, namely that one assigning to i the value 0 [or 1] i.e. $0 \rightarrow i$ [or $1 \rightarrow i$]. In many cases, of which we will present examples, it is necessary to substitute with still more freedom, e.g. $f(i, j, k, \ldots) \longrightarrow i$, where f is a function of i itself as well as of other variables j, k, ... (For details cf. the first part of 7.7.) For this reason we will denote each area in which a coherent change or group of such changes takes place, by a special box, which we call a *substitution box*. Next we consider the changes, actually limitations, of the domains of variability of one or more bound variables, individually or in their interrelationships. It may be true, that whenever C actually reaches a certain point in the flow diagram, one or more bound variables will necessarily possess certain specified values, or possess certain properties , or satisfy certain relations with each other. Furthermore, we may, at such a point, indicate the validity of these limitations. For this reason we will denote each area in which the validity of such limitations is being asserted, by a special box, which we call an *assertion box*.

The boxes of these two categories (substitutions and assertions) are, like the operation boxes, insertions into pieces of the flow diagram where no branchings or mergers take place. They have therefore one input arrow and one output arrow. This necessitates some special marks to distinguish them from operation boxes. In drawing these boxes in detail, certain distinguishing marks of this type will arise automatically (cf. as above), but in order to clarify matters we will also mark every substitution and assertion by a cross #. In addition, we will, for the time being, also mark substitution boxes with an s, followed by the bound variable that is being changed and assertion boxes with an a.

Thus the operation boxes, the substitution boxes and the assertion boxes produce together the dissection of the flow diagram into its constancy intervals. We will number the constancy intervals with Arabic numerals, with the rules for sequencing, subdividing, modifying and correcting as given in Note I in 7.4.

To conclude, we observe that in the course of circling an induction loop, at least one variable (the induction variable) must change, and that this variable must be given its initial value upon entering the loop. Hence the junction before the alternative box of an induction loop must be preceded by substitution boxes along both paths that lead to it: Along the loop and along the path that leads to the loop. At the exit from an induction loop the induction variable usually has a (final) value which is known in advance, or for which at any rate a mathematical symbol has been introduced. This amounts to a restriction of the domain of variability of the induction variable, once the exit has been crossed — indeed, it is restricted from then on to its final value, i.e. to only one value. Hence this is usually the place for an assertion box.

A scheme exhibiting all the features discussed in this section is shown in Figure 7.7.

FIGURE 7.7

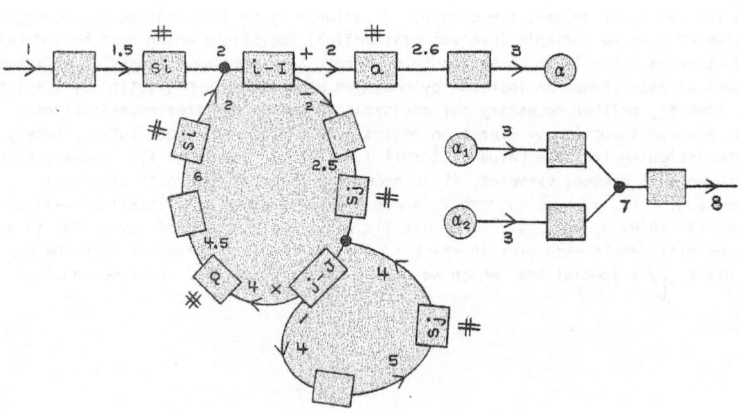

7.7 We now complete the description of our method to indicate the successive stages and the functioning of storage.

The areas in the selectron memory that are used in the course of the procedure under consideration will be designated by capital letters, with the rules for sequencing, subdividing, modifying and correcting as given in Note 1 in 7.4.

The changes in the contents of this storage are effected by the operation boxes, as described in 7.6. An operation box should therefore contain the following indications: First, the expressions which have to be calculated. Second, the storage positions to which these expressions have to be sent subsequently. The latter are stated by an affix "to ....". E.g. if the expression $\sqrt{xy+z}$ (x, y, z are variables or other expressions) is to be formed and then stored at C.2, then the indication will be "$\sqrt{xy+z}$ to C.2". It may also be desired to introduce a new symbol for this expression, i.e. to define one, e.g. $w = \sqrt{xy+z}$ Then the indication will be "$w = \sqrt{xy+z}$ to C.2".

It should be added that an expression that has been calculated in an operation box may also be sent to a variable remote connection (i.e. to its entrance, cf. the last part of 7.4). The expression must, of course, be the designation of one of the exits of that connection, and the operation consists of equating the designation of the entrance to it. The latter is the Greek letter (possibly with decimals, cf. Note 1 in 7.4), the former is this Greek letter with a definite index. No affix "to ...." is needed in this case, since the designation of the entrance determines the connection that is meant. Hence such an operation looks like this: "$\alpha = \alpha_2$" . One or more indications of this kind make up the contents of every operation box.

NOTE 1: An affix "to ...." may also refer to several positions. E.g. "to A, C.1, 2".

NOTE 2: The optional procedure of defining a new symbol by an expression formed in an operation box becomes mandatory when that box is followed by a point of confluence of several paths. Indeed, consider the two operation boxes from which the constancy interval 7 issues in Figure 7.7. Assume that their contents are intended to be "x + y to C.1" and "x - y to C.1", respectively. In order that C.1 have a fixed content throughout the constancy interval 7, as required by our definitions, it is necessary to give the alternative expressions x + y and x - y a common name, say z. Then the contents of the two operation boxes in question will be "z = x + y to C.1" and "z = x - y to C.1".

NOTE 3: A variant of the procedure of defining a new symbol is to define an expression involving a new symbol, which will afterwards only occur as a whole, or to define the value of a function for a new value of its variable (usually in an induction). E.g. "$f(i+1) = \frac{1}{2} f(i) (1 - f(i))$ to A.2".

The contents of a substitution box are one or more substitutions of the type described in 7.6, in connection with the definition of the concept of a substitution box. Such a substitution is written like this: "$f(i, j, k, \ldots) \rightarrow i$". Here i is a bound variable, the one whose value is being changed (substituted), while $f = f(i, j, k, \ldots)$ is an expression which may or may not contain i itself as well as any other bound variables j, k, ... (and any free variables or constants).

The contents of an assertion box are one or more relations. These may be equalities, or inequalities, or any other logical expressions.

The successive contents of the storage positions referred to above (A, B, C, ... with decimal fractions) must be shown in a separate table, the *storage table*. In conjunction with this table a list of the free variables should be given (in one line) and a list of the bound variables (in the next line).

The storage table is a double entry table with its lines corresponding to the storage positions, and its columns to the constancy intervals. Columns and lines are to be marked with the symbols and numbers of the entities to which they correspond. Each field of this table should show the contents of the position that corresponds to its line throughout the constancy interval corresponding to its column. These contents are expressions which may contain bound variables as well as free variables and constants. It should be noted that the bound variables must appear as such, without any attempt to substitute for them their actual values, since only the former are attached to a given constancy interval as such, while the latter depend also on the specific stage in the course of C, in which C passes at a particular occasion through that constancy interval.

It is obviously convenient to fill in a field in the storage table only if it represents a change from the preceding constancy interval. Therefore we reserve the right (without making it an obligation) to leave a field empty if it represents no such change. In this case the field must be thought of as having the same content as its line had in the column (or columns) of the constancy interval (or intervals) immediately preceding it along the flow diagram. (In the plural case, which occurs at a Junction of the flow diagram, these columns [constancy intervals] must all have the same content as the line in question. Indeed, without observing this rule, it would not be possible to treat all incoming branches of a junction as one constancy interval. Cf., e.g. the constancy intervals 2 and 7 in Figure 7.7.) If this antecedent column is not the one immediately to the left of the column in question, it may be advisable to indicate its number in brackets in the field in question.

Certain storage positions (lines) possess no relevance during certain constancy intervals (columns) i.e. their contents are changed or produced and become significant at other stages of the procedure (in other columns). The corresponding field may then be marked with a dash. The repetition rules given above apply to these fields with dashes too.

A scheme exhibiting all these features is shown in Figure 7.8.

FIGURE 7.8

		1	1.5	2	2.5	2.6	3	4	4.5	5	6
I,J	A.1	–	1	i		I	–	[2]			i + 1
i,j	2	–	1	g(i)		g(I)	p	[2]			g(i+1)
	B.1	–			1		–	j	J	j + 1	–
	2	–			½		–	f(j,i)	f(J,i)	f(j+1,i)	–

In certain problems it is advantageous to represent several storage positions, i.e. several lines, by one line which must then be marked by a generic variable, or by an expression containing one or more such variables, and not by a definite number. This variable will replace one or more (or all) of the decimal digits (or rather numbers) following after the capital letter that denotes the storage area in which this occurs, or these variables may enter into an expression replacing those digits (numbers). In a line which is marked in this way the fields will also be occupied by expressions containing that generic variable or variables. "Expressions" include, of course, also explanations which state what expression should be formed depending on various alternative possibilities.

It may be desirable to use such variable-marked lines only in a part of the storage table, i.e. only for certain columns (constancy intervals). Or, one may want to use different systems of variable markings in different parts. This necessitates breaking the storage table up accordingly and grouping those columns (constancy intervals) together for which the same system of variable markings is being used

Since the flow diagram is fixed and explicitly drawn, therefore, the columns of the storage table are fixed, too, and there can be no question of variable markings for them.

In actual practice it may be preferable to distribute all or part of this table over the flow diagram. This can be done by attaching every field at which a change takes place to that constancy interval to which it (i.e. its column) belongs, with an indication of the storage position that it represents. (This applies to fields whose line has fixed markings, as well as to those whose line has variable markings, cf. above.) Since the immediately preceding constancy intervals are at once seen in this arrangement, there is here no need to give any of the indications suggested by our repetition rules. For mnemotechnical reasons it may be occasionally worthwhile to indicate at a constancy interval the contents of certain fields at which no changes are taking place at that stage (e.g. because no change took place there over a long run, and the original indication is at a remote point in the flow diagram). These two methods, the *tabular* and the *distributed* indication of the storage, may also be used together, in mixed forms. Thus, if the distributed method is used in principle, there may be constancy intervals at which more fields have to be described than convenient (e.g. for reasons of space). Such an interval may then be marked by an asterisk, and a *partial storage table* made, which covers by its columns only the constancy intervals with asterisks. Figure 7.8 is the tabular form of the storage table of a flow diagram which is part of that one of Figure 7.7 (without the portion along the constancy intervals 7, 8, and part of 3), and which will be given in full in Figure 7.10. The distributed form is shown in Figure 7.9.

FIGURE 7.9

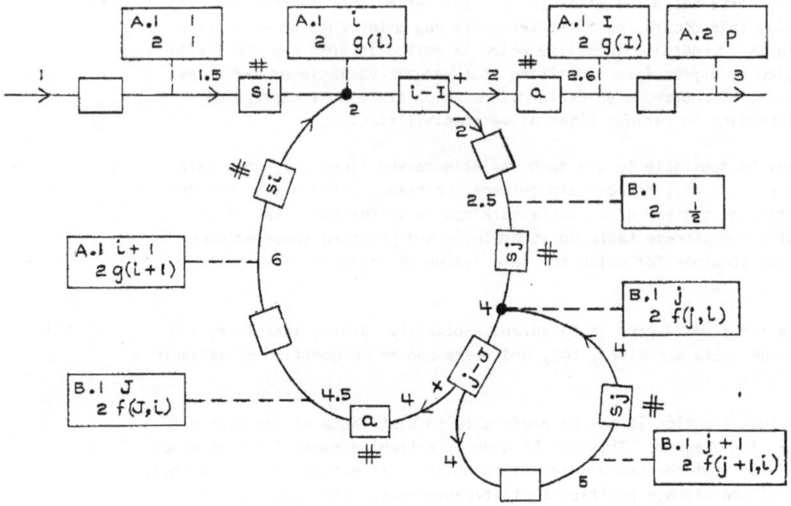

7.8 The only constituents of the flow diagram about which more remains to be said are the operation boxes, the substitution boxes and the assertion boxes — although even for these short descriptions were given in 7.7. We will now discuss them exhaustively.

An operation box contains one or more expressions, and it indicates that these expressions have to be calculated. Every expression may or may not be preceded by a symbol to which it is equated and which it is defining (cf. also Note 3 in 7.7); and it may or may not be followed by an affix "to ....", the .... being occupied by the symbol of a definite position, at which the expressions in question is to be stored. If the affix "to ...." is missing, this means that the expression in question may be stored in the accumulator or the arithmetical register or otherwise, but that at any rate it will not be needed at any remoter future stage.

A substitution box contains one or more expressions of the type "f $\rightarrow$ i", where i is a bound variable, and f is an expression which may or may not contain i and any other bound variables, as well as any free variables and constants. A substitution box is always marked with a cross #.

An assertion box contains one or more relations (cf. the corresponding discussion in 7.7). It is always marked with a cross #.

The three categories of boxes that have just been described, express certain actions which must occur, or situations which must exist, when C passes in its actual course through the regions which they represent. We will call these the *effects* of these boxes, but it must be realized, that these are effects in a symbolic sense only.

The effects in question are as follows:

The bound variables occuring in the expressions of an operation box must be given the values which they possessed in the constancy interval immediately preceding that box, at the stage in the course of C immediately preceding the one at which that box was actually reached. The calculated expression must then be substituted in a storage position corresponding to the immediately following constancy interval, as indicated.

A substitution box never requires that any specific calculation be made, it indicates only what the value of certain bound variables will be from then on. Thus if it contains "f ⟶ i", then it expresses that the value of the bound variable i will be f in the immediately following constancy interval, as well as in all subsequent constancy intervals, which can be reached from there without crossing another substitution box with a "g ⟶ i" in it. The expression f is to be formed with all bound variables in it having those values which they possessed in the constancy interval immediately preceding the box, at the stage in the course of C immediately preceding the one at which that box was actually reached. . (This is particularly significant for i itself, if it occurs in f.)

An assertion box never requires that any specific calculations be made, it indicates only that certain relations are automatically fulfilled whenever C gets to the region which it occupies.

The contents of the various fields of the storage table (tabulated or distributed) i.e. the various storage positions at the various constancy intervals must fulfill certain conditions. It should be noted that these remarks apply to the contents of the field in the flow diagrams, and not to the contents of the corresponding positions in the actual machine at any actual moment. The latter obtain from the former by substituting in each case for every bound variable its value according to the actual stage in the course of C -- and of course for every free variable its value corresponding to the actual problem being solved. Now the conditions referred to above are as follows:

First: The interval in question is immediately preceded by an operation box with an expression in it that is referred "to ...." this field: The field contains the expression in question, unless that expression is preceded by a symbol to which it is equated and which it is defining (cf. also Note 3 in 7.7), in which case it contains that symbol.

Second: The interval in question is immediately preceded by a substitution box containing one or more expressions "f ⟶ i", where i represents any bound variable that occurs in the expression of the field: Replace in the expression of the field every occurrence of every such i by its f. This must produce the expression which is valid in the field of the same storage position at the constancy interval immediately preceding this substitution box.

Third: The interval in question is immediately preceded by an assertion box: It must be demonstrable, that the expression of the field is, by virtue of the relations that are validated by this assertion box, equal to the expression which is valid in the field of the same storage position at the constancy interval immediately preceding this assertion box. If this demonstration is not completely obvious, then it is desirable to give indications as to its nature: The main stages of the proof may be included as assertions in the assertion box, or some reference to the place where the proof can be found may be made either in the assertion box or in the field under consideration.

Fourth: The interval in question is immediately preceded by a box which falls into neither of the three above categories. The field contains a repetition of what the field of the same storage position contained at the constancy interval immediately preceding this box.

Fifth: If the interval in question contains a merger (of several branches of the flow diagram), so that it is immediately preceded by several boxes, belonging to any or all of the four above categories, then the corresponding conditions (as stated above) must hold with respect to each box.

Finally: The contents of a field need not be shown, if it is felt that the omission (or rather the aggregate of simultaneously effective omissions of this type) will not impose a real strain on the reader, due to the amount of implicitly given material that he must remember. Such omissions will be indicated in many cases where mere repetitions of previous contents are involved. (Cf. the remarks made in this connection in 7.7, in the course of the discussion of the distributed form of storage.)

The storage table need not show all the storage positions actually used. The calculations that are required by the expressions of an operation box or of an alternative box may necessitate the use of additional storage space. This space is then specifically attached to that box, i.e. its contents are no longer required and its capacity is available for other use as soon as the instructions of that box have been carried out.

Figure 7.10 shows a complete flow diagram. It differs from that one of Figure 7.9 only inasmuch that the operation boxes and the storage boxes were left empty then. It is easily verified that it represents the (doubly inductive) procedure defining the number p, which is described under it.

FIGURE 7.10

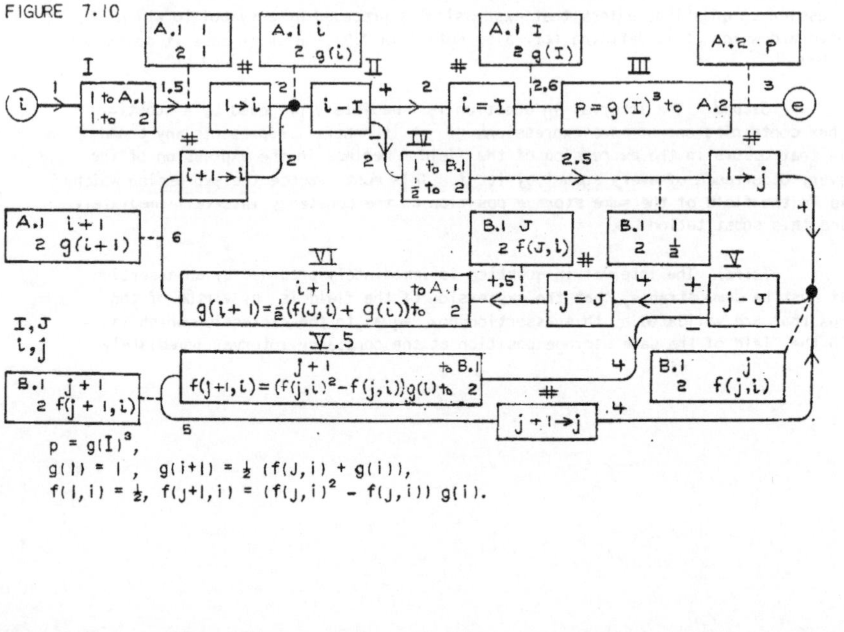

$p = g(I)^3,$

$g(1) = 1, \quad g(i+1) = \frac{1}{2}(f(J,i) + g(i)),$

$f(1,i) = \frac{1}{2}, \quad f(j+1,i) = (f(j,i)^2 - f(j,i))\,g(i).$

Among the boxes shown on this flow diagram the operation boxes and the alternative boxes and the variable remote connections require further, detailed coding. The substitution boxes and the assertion boxes (i.e. the boxes which are marked by a cross #) are purely explanatory, and require no coding, as pointed out earlier in 7.8. The storage boxes have to be coded essentially as they stand. (For all this cf. the details given in 7.9.) Thus the remaining problem of coding is attached to the operation boxes, the alternative boxes and the variable remote connections, and it will prove to be in the main only a process of static translation (cf. the end of 7.1 as well as 7.9). In order to prepare the ground for this final process of detailed (static) coding, we enumerate the operation boxes and the alternative boxes by Roman numerals, with the rules for sequencing, subdividing, modifying and correcting as given in Note 1 in 7.4. Figure 7.10 shows such an enumeration. The variable remote connections are already enumerated.

Finally, we indicate the beginning and the end of the completed flow diagram by two circles (1)→ and →(e), cf. Figure 7.10

7.9 We can now describe the actual process of coding. It is a succession of steps that take place in the following order.

First: Coding is, of course, preceded by a mathematical stage of preparations. The mathematical or mathematical-physical process of understanding the problem, of deciding with what assumptions and what idealizations it is to be cast into equations and conditions, is the first step in this stage. The equations and conditions thus obtained are rigorous, with respect to the system of assumptions and idealizations that has been selected. Next, these equations and conditions, which are usually of an analytical and possibly of an implicit nature, must be replaced by arithmetical and explicit procedures. (These are usually step-by-step processes or successive approximation processes, or processes with both of these characteristics — and they are almost always characterized by multiple inductions.) Thus a procedure obtains, which is approximate in that sense in which the preceding one was rigorous. This is the second step in this stage.

It should be noted that the first step has nothing to do with computing or with machines: It is equally necessary in any effort in mathematics or applied mathematics. Furthermore, the second step has, at least, nothing to do with mechanization: It would be equally necessary if the problems were to be computed "by hand".

Finally, the precision of the approximation process, introduced by the second step, must be estimated. This includes the errors due to the approximations introduced by the second step, as well as the errors due to the machine's necessarily rounding off to a fixed number of digits (in our projected machine to a sign plus 39 binary digits, cf. Part I of this report) after every intermediate operation (specifically: after every multiplication and division). These are, of course, the well-known categories of *truncation errors* and of *round-off errors*, respectively. In close connection with these it is also necessary to estimate the sizes to which the numbers that occur in any part and at any stage of the calculation, may grow. After these limits have been established, it is necessary to arrange the calculation so that every number is represented by a multiple (by a fixed power of 2) which lies in the range in which the machine works (in our projected machine between -1 and 1, cf. Part I of this report). This is the third and last step of this stage. Like the second step, it is necessary because of the computational character of the problem, rather than because of the use of a machine.

In our case there exists an alternative with respect to this third step: It may be carried out by the planner, "mathematically", or it may be set up for computation, in which case it may be advantageous to have it, too, carried out by the machine.

After these preliminaries are completed, the coding proper can begin.

Second: Coding begins with the drawing of the flow diagrams. This is the *dynamic* or *macroscopic* stage of coding. The flow diagram must be drawn on the basis of the rules and principles developed in 7.3 - 7.8. It has been our invariable experience, that once the problem has been understood and prepared in the sense of the preceding first remark, the drawing of the flow diagram presents little difficulty. Every mathematician, or every moderately mathematically trained person should be able to do this in a routine manner, if he has familiarized himself with the main examples that follow in this report, or if he has had some equivalent training in this method.

It is advisable to draw a (usually partial) storage table for the main data of the problem, and use from then on either purely distributed, or mixed distributed and tabular storage, pari passu with the evolution of the diagram. The flow diagram is, of course, best started with one of the lowest (innermost) inductions, proceeding to the higher inductions which contain it, reverting (after these are exhausted) to another lowest induction (if any are left), etc., etc. No difficulty will be found in keeping the developing flow diagram complete at every stage, except for certain enumerations which are better delayed to the end (cf. below).

It is difficult to avoid errors or omissions in any but the simplest problems. However, they should not be frequent, and will in most cases signalize themselves by some inner maladjustment of the diagram, which becomes obvious before the diagram is completed. The flexibility of the system of 7.3 - 7.8 is such that corrections and modifications of this type can almost always be applied at any stage of the process without throwing out of gear the procedure of drawing the diagram, and in particular without creating a necessity of "starting all over again".

The enumeration of the distributed storage and of the remote connections and of the constancy intervals should be done pari passu with the drawing of the diagram. The enumeration of the operation boxes and of the alternative boxes, on the other hand, is best done at the end after the flow diagram has been completed.

After a moderate experience has been acquired, many simplifications and abbreviations will suggest themselves to almost any coder. These are best viewed as individual variants. We wish to mention only one here: If distributed storage is extensively used, only a few (if any) among the constancy intervals will have to be enumerated.

Third: The next stage consists of the individual coding of every operation box, alternative box and variable remote connection. This is the *static* or *microscopic* stage of coding. Its main virtue is, that the boxes in question can now be taken up one by one, and that the work on each one of them is essentially unaffected by the work on the others. (With one exception, cf. below.)

We feel certain that a moderate amount of experience with this stage of coding suffices to remove from it all difficulties, and to make it a perfectly routine operation. The actual procedure will become amply clear by reading the examples that follow in this report. We state here only general principles which govern the procedure.

The coding of a variable remote connection is best attached to the immediately preceding box. If it is preceded by several such boxes, either of these may be used; if no suitable box is available, it may be given a Roman numeral and treated as a separate box.

The coding of the (operation or alternative) boxes remains. The coding of each box is a separate operation, and we called it, as such, static. This is justified in this sense: The course of C during such a period of coding is strictly linear, i.e. there are no jumps forward or backward (by the Cu or Cc type transfer orders — the orders 13-16 in Table II) — except possibly at the end of the period — C moves without omitting anything and without ever going twice over the same ground (within one period of this type). It is in this sense only that the process is static: Substitutions, i.e. changes in the memory (hence either in the storage or in the orders), occur at every step. In addition, partial substitutions (by the Sp type substitution orders — the orders 18,19 in Table II) cause a slight deviation from strict linearity: They affect orders (the ones which are to be modified by substitution) which may not have been coded yet. In this situation it is best to leave in the substitution order the space reserved for the position mark of the order to be substituted (in the preliminary enumeration, cf. below) empty, and mark the substitution order as incomplete. After all orders have been coded, all these vacancies can be filled in. These filling-in operations can be effected within a single linear passage over the entire coded sequence.

The coding of such a box occurs accordingly as a simple linear sequence, and it is therefore necessary to define a system for the enumeration of the code orders within the sequence (of this box). We call it the *preliminary enumeration*, because the numbers which we assign at this stage are not those that will be the x's of the actual position in the selectron memory. The latter form the *final enumeration* to be discussed further below. The preliminary enumeration is defined as follows: The symbol of the box under consideration is to be shown, then a comma, and then an Arabic numeral — the latter being used to enumerate the code orders of this sequence. (We might again allow the rules for sequencing, subdividing, modifying and correcting as given in Note I in 7.4, for this enumeration, too. It will, however, hardly ever be necessary to have recourse to these — there will almost never be any difficulty in carrying out a simple linear numbering of the entire sequence pari passu with its coding.)

Thus the enumeration for a box II.I might be II.I, I; II.I, 2; ... . Actually these symbols will be written under each other, and the box symbol (II.I in the above case) need be shown only the first time. In every order the expression S(x) will have to be written with the full preliminary symbol of x, e.g.; S(II.I,3). (This example must, of course, be thought of as referring to a selectron position x containing an order. If x contained a number, an expression like S(A.I,2) would be typical.)

References to a variable remote connection are best made by its Greek letter (possibly with decimals, cf. Note I in 7.4, and with or without indexing according to whether the entrance or the exit is meant), since it may not be feasible or convenient to determine at this stage the identifications of its parts with the appropriate (Roman numeral) boxes.

Any storage that becomes necessary in the course of this coding (in excess of the tabulated or distributed storage of the flow diagram, i.e. the storage attached to the box, cf. the last part of 7.8), has to be shown separately, as *local storage*. The local storage should be enumerated in the same way as indicated above, but with a symbol s after the comma, e.g. : II.1, s.1; II.1, s.2; ... (For an S(x) which refers to such a position one might write, e.g. S(II.1, s.2), in the sense of the suggestion in the preceding paragraph. However, since the reference is by its nature to the box under consideration at the time, it is legitimate to abbreviate by omitting the box symbol, writing e.g. S(s.2).)

The coding itself should be done in a column of its own, enumerated as indicated above. It is advisable to parallel it with another, explanatory, column, which shows for each order the effect of the substitution that it causes. This effect appears, of course, in one of the following places: The accumulator (symbol: Ac), or the arithmetical register (symbol: R), or some storage position or coded order (symbols as described up to now). The explanatory column then shows each time the symbol of the place where the substitution takes place, and its contents after the substitution. The Cu or Cc type transfer orders (the orders 13-16 in Table I) cause no substitution, they transfer C instead (and they occur only at the end of the sequence) — hence they require no explanation.

An order in the sequence, which becomes effective in a substituted form (substituted by an earlier order in the sequence), should be shown in its original form and then, in brackets, in its substituted form. The original form may have a –, the sign of irrelevance, or any other symbol (e.g. the number 0, of course, as a 12 digit binary), in place of its x (in its S(x)). It is this form which matters in the actual coding. (In writing out a code, it is best to use the sign of irrelevance in such a case. When it comes to real coding, however, an actual number must be used, e.g. 0.) The substituted form, on the other hand, is clearly the one which will actually control the functioning of the machine, and it is the one to which the explanation should be attached.

We repeat: We feel certain that direct, linear coding according to this principle presents no difficulties after a moderate amount of experience has been acquired.

Fourth: The last stage of coding consists of assigning all storage positions and all orders their final numbers. The former may be enumerated in any order, say linearly. The latter may then follow, also in a linear order, but with the following restriction: The flow diagram indicates that a certain box must follow immediately after another operation box, or immediately after the – branch of an alternative box. (– is the case where the transfer of C by a conditional transfer order — type Cc — does not become effective). In this case the sequence of the former box must begin immediately following the end of the sequence of the latter box. It can happen, that this principle requires that a box be the immediate successor of several boxes. In this case all of the latter boxes (except one) must be terminated by (unconditional) transfer orders to the former box. Such a transfer order is also necessary if a box requires as its immediate successor a non-initial order of another box.

At this stage the identification of every part (entrance and exits) of every variable remote connection with appropriate (Roman numeral) boxes (i.e. orders in them) must be established. The Greek letter references to these connections will then be rewritten accordingly.

These principles express all the restrictions that need be observed in the final, linear ordering of the boxes. To conclude, we note that it must be remembered, that two orders constitute together a word, i.e. that they have together one final number, and are distinguished from each other as its "left" or "right" order.

After this final enumeration has been completed, one more task remains: In every order, in the expression S(x) the preliminary number x must be replaced by the final number x, and for those orders which substitute positions that are occupied by orders (orders without or with primes -- these are the orders 13-16 and 18, 19 in Table II) the distinction between "left" and "right" must be made (i,e. the order must not be primed or it must be primed, respectively). In the case of positions which correspond to the exits of one variable remote connection it is clearly necessary that they be all "left" or all "right". This must be secured; the ways to achieve this are obvious.

When the coding is thus completed, the positions where the sequence of orders begins and ends (corresponding to i and e, cf. the end of 7.8), should be noted. There will be more to say about these when we get to the questions of subroutines and of combining routines, to which reference will be made further below, but we will not go into this matter now.

All these things being understood, it is indicated to state this in addition: There are still some things which have to be discussed in connection with the coding of problems, and which we have neglected to consider in this chapter, and will also disregard in the chapters which follow in this Part II of the report. We refer to the orders which stop the machine, to the orders which control the magnetic wire or tape and the oscilloscopic output of the machine, and to the logical principles and practical methods applying to the use of the magnetic wire or tape inputs and outputs as a subsidiary memory of the machine.

We made brief references to these matters in the paragraphs 4.5, 4.8, 6.8 (and the sub-paragraphs of the latter) in Part I of this report. They require, of course, a much more detailed consideration. That we neglect to take them up here, in Part II of our report is nevertheless deliberate. The main reason for this neglect is, that they depend on parts of the machine where a number of decisions are still open. These are decisions which lead only to minor engineering problems either way, but they do affect the treatment of the three subjects mentioned above (stop orders, output orders, use of the input-output as a subsidiary memory) essentially. So the corresponding discussion is better withheld for a later report.

Furthermore, the two first items are so simple, that they do not seriously impair the picture that we are going to present in what follows. The potentialities of the third item are much more serious, but we do not know as yet, to what an extent such a feature will be part of the first model of our machine.

There are further important general principles affecting the efficient use of coding and of routines. These are primarily dealing with *general routines* and *sub-routines* and with their use in *combining routines* to new routines. These matters will be taken up, as far as the underlying principles are concerned in Chapter 12, and put to practical use in various examples, with further discussion in the subsequent Chapters 13 and 14. In the immediately following Chapters 8 - 11, we will give examples of coding individual routines, on the basis of the general principles of this Chapter 7.

8.0 CODING OF TYPICAL ELEMENTARY PROBLEMS.

8.1 In this chapter, as well as in the following ones, we will apply the general principles of coding, as outlined in Chapter 7, to a sequence of problems of gradually increasing complexity. The selection of these problems was made primarily with the view of presenting a typical selection of examples, on which the main problems, difficulties, procedures and simplifying and labor saving methods or tricks of the actual coding process can be exhibited and studied. They will be chosen from various parts of mathematics, and some of them will be of a more logical than mathematical character. After having progressed from simple problems to complicated ones within the category of those problems which have to be coded as units, i.e. without breaking up into smaller problems, we will proceed to problems which can be broken up into parts, that can be coded separately and then fitted together to code the whole problem.

8.2 Before we start on this program of coding specific problems, however, there remains one technical point that has to be settled first.

In all the coding that we are going to do, references to specific memory locations will occur. These locations are 2^{12}=4,096 in number, and we always enumerated them with the help of a 12 binary digit number or aggregate, which we usually denoted by x. (Cf., e.g. through the Table II of orders.) We will use for this 12 binary digit number or aggregate, in all situations where it may occur, the generic designation of a *memory position mark*, or briefly *position mark*. By calling it ambiguously number, aggregate or mark, we have purposely avoided the question whether and how we wish to attribute a binary point position to this digital aggregate. We now make the choice in the most convenient way: Let the binary point be at the extreme right, i.e. let the position marks be integers. Thus the position marks correspond to all 12 binary digit integers, i.e. to all (decimal) integers from 0 to $2^{12}-1$ = 4,095.

This choice, however, is not entirely without consequences for the way in which the position marks themselves must be placed into the memory, and manipulated arithmetically by the machine, whenever such operations are called for. The point is that the machine will not handle integers as such, but only numbers between -1 and 1. Furthermore, in order that a position mark x that is stored in the memory should become effective, it is necessary to substitute it into some order. This involves bringing it from its storage position into the accumulator (by an order S(y)⟶ Ac + [abbreviated: y], assuming that the storage position in question has the mark y), and then make a partial substitution from the accumulator into the desired order (by an order Ap⟶ S(z) or Ap' ⟶ S(z) [abbreviated: zSp or zSp'], assuming that the order to be substituted has the position mark z).

If we followed the explanation of these orders (Nos. 18, 19) given in Table I or in 6.6.5 in Part I of this report, then both orders would move the left-hand 12 digits of the accumulator into the proper places at z (into the left-hand 12 digits of the left-hand or right-hand order there, i.e. into digits 1 to 12 or 21 to 32 [from the left] in that word). Now the position mark to be moved, x, is according to the convention that we have chosen, an integer. Digit 1 in the accumulator is the sign digit, while digit i (= 2, ..., 40) there has the positional value $2^{-(i-1)}$.

Therefore the accumulator should actually contain not x itself, but $2^{-11}x$ (modulo 2). Hence x should be stored in the memory (at y, cf. above) as $2^{-11}x$ (modulo 2).

From an engineering point of view it is preferable not to involve the sign digit of the accumulator into these transfers, and not to effect both transfers (into the left-hand orders and into the right-hand orders) from the same 12 accumulator digits. This motivates the change which we made in Table II, as against Table I, for these orders 18, 19: We effect these two transfers from the digits 9-20 and 29-40 (from the left) of the accumulator, respectively.

In view of these charges it is desirable to have in the accumulator the 12 digits of x both at the digits 9 to 20, and at the digits 29 to 40, when the substitutions of the type under discussion are being effected. I.e. the accumulator should contain $2^{-19}x + 2^{-39}x$.

We define accordingly:

A(memory) position mark x is always a 12 binary digit integer, i.e. a (decimal) integer from 0 to $2^{12} - 1 = 4{,}095$. Form

$$x_0 = 2^{-19}x + 2^{-39}x.$$

Then whenever x has to be stored in the memory, or in any manner operated on by the machine, it has to be replaced by the number x_0 given above.

The symbols introduced in Chapter 7 to denote storage positions for the tabular or distributed storage (English capitals with decimals, cf. the beginning of 7.7) or the local storage (Roman numeral with decimals [denoting operation or alternative boxes] followed by an s and a decimal, cf. the latter part of 7.9) or for remote connections (Greek letters with decimals, cf. Note I in 7.4) will be considered to be position marks, i.e. integers x, and treated accordingly, in the sense of the above rule.

8.3 We now take up our first actual coding problem. This is a very elementary problem, of a purely algebraical character and presenting no logical complications whatever.

Problem 1.

The constants a, b, c, d, e and the variable u are stored in certain, given, memory locations. It is desired to form the expression

$$v = \frac{a\,u^2 + b\,u + c}{d\,u + e}$$

and to store it at another, given, memory location. ——

The problem is so simple that it can be coded directly, without a flow diagram. We may, therefore, attribute all storage to a single storage area A, and all operational orders to a single operation box I. (Both of these are, in a way, "fictitious", since we do not draw a flow diagram.)

Accordingly, let the constants a, b, c, d, e be stored at A.1 to 5, the variable u at A.6, the (desired) quantity v at A.7. Note that u is a free variable, a to e are constants which may be viewed as free variables, while v is produced by the machine itself and the content of A.7 at the start is therefore irrelevant.

Obviously the expressions that will have to be formed successively are:

1) au, au + b, au^2 + bu = (au + b) u, au^2 + bu + c.

2) du, du + e.

3) $$v = \frac{au^2 + bu + c}{du + e}$$

Intermediate, i.e. local, storage is needed for whichever of the two quantities au^2 + bu + c and du + e is computed first, while the second one is obtained. However, at this stage A.7 is still available (v is not formed) and may be utilized for that storage. It is convenient when the division of 3) begins to have its dividend au^2 + bu + c in the accumulator. Hence 1) should immediately precede 3). Therefore 1) to 3) should be taken up in the order 2), 1), 3).

Some additional local storage is required for a certain quantity in transit, at a moment when A.7 is no longer available. (This quantity is au + b, which has to move from Ac to R, while A.7 is occupied by du + c. Cf. the orders 1.8,9 below.) This requires an additional storage position A.8.

The entire coding is static, and may be arranged as follows:

Storage at the beginning:

A.1	a	A.4	d	A.7	--
2	b	5	e	8	--
3	c	6	u		

Coded sequence (preliminary enumeration), detailed form:

	(Coding Column.)	(Explanatory Column.)	
1,1	S(A.4) → R	R	d
2	S(A.6) xR → A	Ac	du
3	S(A.5) → Ah +	Ac	du + e
4	At → S(A.7)	A.7	du + e
5	S(A.1) → R	R	a
6	S(A.6) xR → A	Ac	au
7	S(A.2) → Ah+	Ac	au + b
8	At → S(A.8)	A.8	au + b
9	S(A.8) → R	R	au + b
10	S(A.6) xR → A	Ac	au^2 + bu
11	S(A.3) → Ah +	Ac	au^2 + bu + c
12	A ÷ S(A.7) → R	R	$v = \frac{au^2 + bu + c}{du + e}$
13	R → A	Ac	
14	At → S(A.7)	A.7	

Note, that I,1-4 corresponds to 2) above, I,5-11 to 1), and I,12-14 to 3). In all the codings which follow we will use the abbreviated notation of the Table II of orders. Then the above coded sequence (preliminary enumeration) assumes the following appearance:

I,1	A.4	R	R	d
2	A.6	x	Ac	du
3	A.5	h	Ac	du + e
4	A.7	S	A.7	du + e
5	A.1	R	R	a
6	A.6	x	Ac	au
7	A.2	h	Ac	au + b
8	A.8	S	A.8	au + b
9	A.8	R	R	au + b
10	A.6	x	Ac	$au^2 + bu$
11	A.3	h	Ac	$au^2 + bu + c$
12	A.7	÷	R	$v = \dfrac{au^2 + bu + c}{du + e}$
13		A	Ac	
14	A.7	S	A.7	

The passage to the final enumeration consists of assigning A.1-8 their actual values, and of pairing the 14 orders I,1-14 to 7 words. Let the latter be 0 to 6, and let A.1-8 be 7 to 14. Then the coded sequence becomes (the explanatory column is no longer needed):

0	10 R , 12 x	5	9 h , 13 ÷	10	d
1	11 h , 13 S	6	A , 13 S	11	e
2	7 R , 12 x	7	a	12	u
3	8 h , 14 S	8	b	13	-
4	14 R , 12 x	9	c	14	-

In an actual application of this coded sequence the constants (or parameters, or free variables) a to e and u, occupying the locations 7, .., 12, would, of course, have to be replaced by their actual, numerical values. The other locations, 1, ..., 6, on the other hand, will have to be coded exactly as they stand.

These remarks (concerning constants, or parameters, or free variables) apply, of course, equally to all coded sequences that we will set up in the remainder of this report. (Cf., however, the remarks on subroutines and on combining routines in Chapter 12, which have a certain significance in this respect.)

8.4 We introduce next a slight logical twist into Problem 1.

Problem 2.

Same as Problem 1, with this change: u and v are stored or to be stored at certain locations m and n, which are not given in advance, but stored at certain, given, memory locations. ---

This affects the use of storage. A flow diagram is still not necessary, but we must reconsider the use of A.1-8.

A.6,7 are no longer needed for u, v, however, we still need A.7 for local storage (cf. 8.3). Let m, n be stored at A.9,10, of course, in the form m_o, n_o (cf. 8.2).

When we will pass from the preliminary enumeration to the final one, the following observations will apply: First, A.6 is missing, but this is irrelevant. Second, the positions m, n are supposed to be part of some other routine, already in the machine, therefore they need not concern us at this stage. Third, let us assume that the coded sequence is supposed to begin at some position other than 0, say 100, and that it is supposed to end at a prescribed position, say 50.

The abbreviated, preliminary coded sequence 1,1-14 will therefore have to be modified in two ways: First, the references to A.6 (which transfer u) and to A.7 (inasmuch as they transfer v, and are not due to the use of A.7 for local storage) have to be changed. Since u is needed repeatedly, it is advisable to assign to it a fixed position A.11. Hence 1,2, 6, 10 will have to refer to A.11 instead of A.6, and new orders will be needed to get u from m to A.11, and (replacing 1,14) to get v from Ac to n. Second, a new order is needed at the end of the coded sequence to get the control to the desired endpoint 50.

The changes in 1,2, 6, 10 are trivial.

The transfer of u may take place at any time before 1,2. 1,2 assumes that d is in R, so this transfer must either precede 1,1, too (since this moves d to R), or leave R undisturbed. R is indeed not disturbed but we place the transfer routine nevertheless before 1,1. It can be coded as follows:

1,0.1	A.9		Ac	m_o
.2	1,0.3	Sp	1,0.3	m
.3	-			
[	m	]	Ac	u
.4	A.11	S	A.11	u

Note that the sign of irrelevance in an order, like 1,0.3 above, does not mean that the entire order, i.e. the entire half-word is irrelevant (and still less, that the entire word is irrelevant). It means that an order S(x) ⟶ Ac, abbreviated x, is needed, but that x is irrelevant.

The transfer of v replaces the order 1,14. 1,14 assumes that v is in Ac, so the operations required by this transfer must either precede 1,1-14 (since all of these are needed to produce v in Ac), or leave Ac undisturbed. Ac is disturbed, so it is necessary to place the transfer routine before 1,1. It is simplest to let it follow immediately after our first transfer routine, ie. immediately after 1,0.4 above. It can be coded as follows:

1,0.5	A.10		Ac	n_o
.6	1,14	Sp	1,14	n

1,14, in turn, must be rewritten as follows:

1,14		S		
[		S]		n

              Finally, we need an order after 1,14, to move the control to the desired endpoint 50. This is, of course:

1,15	50	

              We can now rewrite the abbreviated, preliminary coded sequence:

Storage at the beginning:

m	u	A.3	c	A.8	–
n	v	4	d	9	m_o
A.1	a	5	e	10	n_o
2	b	7	–	11	–

Coded sequence (preliminary enumeration):

1,0.1	A.9		Ac	m_o
.2	1,0.3	Sp	1,0.3	m
.3	–			
[	m	]	Ac	u
.4	A.11	S	A.11	u
.5	A.10		Ac	n_o
.6	1,14	Sp	1,14	n
1	A.4	R	R	d
2	A.11	x	Ac	du
3	A.5	h	Ac	du + e
4	A.7	S	A.7	du + e
5	A.1	R	R	a
6	A.11	x	Ac	au
7	A.2	h	Ac	au + b
8	A.8	S	A.8	au + b
9	A.8	R	R	au + b
10	A.11	x	Ac	$au^2 + bu$
11	A.3	h	Ac	$au^2 + bu + c$
12	A.7	÷	R	$v = \dfrac{au^2 + bu + c}{du + e}$
13		A	Ac	
14	–	S		
[	n	S }		n
15	50	C		

              The passage to the final enumeration consists of assigning A.1-5, 7-11 their actual values, pairing the 21 orders 1,0.1-15 to 11 words, and then assigning 1,0.1-15 their actual values. We treat m, n like a to e or u: As free variables or constants. The 11 words in question must begin with 100, so they will be 100 to 110. Then A.1-5 may be 111 to 115, and A.7-11 may be 116-120. So we obtain this coded sequence:

100	118 ,	101 Sp	107	117 R ,	120 x	114	d
101	- ,	120 S	108	113 h ,	116 ÷	115	e
102	119 ,	109 Sp'	109	A ,	- S	116	-
103	114 R,	120 x	110	50 C ,	- -	117	-
104	115 h,	116 S	111	a		118	m_o
105	111 R,	120 x	112	b		119	n_o
106	112 h,	117 S	113	c		120	-

Note, that the sign of irrelevance in an order, like 101-left, means that it is an order x (i.e. S(x) ⟶ Ac) with x arbitrary. This is exactly the same situation as in 109-right, which is -S, i.e. x S (i.e. At ⟶ S(x)) with x irrelevant. On the other hand, 110-right, marked - -, is actually entirely irrelevant, since this half-word was not needed for any of our orders. Finally the sign of irrelevance in a full word, like 116 or 117 or 120, means, of course, that the entire content is irrelevant. The simplest treatment of all these irrelevancies is to replace anything that is irrelevant by 0

8.5 The coding of any problem should be followed by an estimate of the time that the machine will consume in solving the problem, i.e. in carrying out the instructions of the coded sequence.

In order to make such an estimate it is necessary to have some information about the duration of each step that is involved, i.e. the time required by the machine to carry out each one of the orders of Table II. At the present stage of our engineering development the order-time estimates are necessarily guesses. The estimates which follow are made in a guardedly optimistic spirit. Unless there are some major surprises in our engineering developments, they should be correct within a moderate error-factor -- say, within a factor 2.

In this sense our order-time estimates are as follows:

TABLE III

1	x	25 μ	8	xh-M	35 μ	15	xCc	30 μ
2	x-	25	9	xR	25	16	xCc'	30
3	xM	30	10	A	5	17	xS	25
4	x-M	30	11	xX	100	18	xSp	25
5	xh	30	12	x÷	150	19	xSp'	25
6	xh-	30	13	xC	25	20	L	5
7	xhM	35	14	xC'	25	21	R	5

plus 20 μ for every word.

1 μ = 1 microsecond, 1 m = 1 millisecond.

Considering the considerable uncertainties with which these estimates are affected, we might as well use the following approximate rule:

Count 75 μ for every word which contains two orders, subtract 25 μ if only one order is used, subtract 20 μ for every order which refers to no memory position x (i.e. A, R, L), add 70 μ for every multiplication and 120 μ for every division.

On this basis the coded sequences obtained in 8.3 and 8.4 should require

$$(7 \times 75 - 20 + 3 \times 70 + 120)\ \mu = 835\ \mu \approx .8\ m,$$

and

$$(11 \times 75 - 25 - 20 + 3 \times 70 + 120)\ \mu = 1{,}110\ \mu \approx 1.1\ m,$$

respectively.

8.6 The next point to consider in connection with our coding of the Problems 1 and 2 is this:

We observed repeatedly that our machine will only deal with numbers lying between -1 and 1. Ordinarily we will interpret this to exclude both endpoints, i.e. we will keep all absolute values < 1. It should be noted, however, that the precise definition of the numbers that our machine can handle as digital aggregates (cf. 5.7 in Part 1 of this report) includes -1 and excludes 1.

This fact, that we have a -1 but not a 1, is occasionally of importance. Actually this is not often the case, indeed extremely less frequently than one might think. There is hardly ever a need to store a 1: Its effect in multiplying or dividing is nil, and the sizes which we wish to observe make it usually inconvenient to manipulate. (Regarding position marks, where 1 might play a role, cf. 8.2. Regarding induction indices, cf. Problem 3 in 8.7 as well as various subsequent Problems.) In those exceptional cases where 1 is actually arithmetically needed, it is always possible to adjust the operations so that -1 can be used instead. (For an example of this cf. Problem 9 in 9.9, in particular the treatment of the box III -- C.2 and III,1 — as well as similar situations in the boxes VI, IX, XII, XV.) The complications which arise in this manner are very rare, and when they occur, negligible, as the experience of our Problems amply illustrates.

In the present case, we will keep all absolute values < 1. It is in this sense, therefore, that we observe that the above problems can only be handled if their u and v lie between -1 and 1. Furthermore, our procedure of 8.3 (and 8.4) also requires that all data of the problem, i.e. the constants a to e should lie between -1 and 1, as well as all intermediate results of the calculation, i.e. the expressions

1) au, $au + b$, $au^2 + bu$, $au^2 + bu + c$,

2) du, $du + e$.

Now assume that u is only known to lie between -2^p and 2^p, v between -2^q and 2^q, and a to e between -2^r and 2^r ($p, q, r = 0, 1, 2, \ldots$). Then the quantities a to e and u, v simply cannot be stored by the machine, and it is necessary to store others instead, e.g.

$$a' = 2^{-r}a, \ldots, \quad e' = 2^{-r}e, \quad u' = 2^{-p}u, \quad v' = 2^{-q}v.$$

This, however, is not an answer to our problem since the relationship between u' and v' is not expressed with the help of a' to e' in the same manner as the relationship between u and v with the help of a to e. Hence we have to proceed somewhat more circumspectly.

This is not difficult. Our original relation

$$v = \frac{au^2 + bu + e}{du + e}$$

implies

$$v' = \frac{a^* u'^2 + b^* u' + c^*}{d^* u' + e^*}$$

where

$$u' = 2^{-p}u, \quad v' = 2^{-q}v,$$

if

$$a^* = 2^{2p-q-s}a, \quad b^* = 2^{p-q-s}b, \quad c^* = 2^{-q-s}c, \quad d^* = 2^{p-s}d, \quad e^* = 2^{-s}e,$$

where $s = 0, \pm 1, \pm 2, \ldots$ can be freely chosen. We can now use our freedom in choosing s to force a^* to e^* as well as the expressions

$$1^*) \quad a^* u', \ a^* u' + b^*, \ a^* u'^2 + b^* u', \ a^* u'^2 + b^* u' + c^*,$$

$$2^*) \quad d^* u', \ d^* u' + e^*$$

into the interval $-1, 1$. Since

$$\left| \frac{a^* u'^2 + b^* u' + c^*}{d^* u' + e^*} \right| = \left| v' \right| < 1,$$

we may disregard the last expression in 1^*), if all others are taken care of. Since $|u'| < 1$, it suffices to secure

$$\left|a^*\right|, \left|b^*\right|, \left|d^*\right|, \left|e^*\right| < \tfrac{1}{2}, \left|c^*\right| < 1.$$

This means that

$$s \geqq 2p-q+r+1, \ p-q+r+1, \ -q+r, \ p+r+1, \ r+1.$$

Hence we may choose

$$s = \text{Max}\,(2p - q + r + 1, \ p + r + 1) = p + \text{Max}\,(p - q, 0) + r + 1.$$

This procedure is typical of what has to be done in more complicated situations. It calls for some remarks, which we proceed to formulate.

First: a to e and u are data, and it is therefore natural that we have to assume that some information, which is separate from this calculation, exists regarding their sizes. (I.e. that they lie between -2^r and 2^r, and between -2^p and 2^p, respectively.) v, however, is determined by a to e and u, so its limits (-2^q and 2^q) should be derivable from the information that exists concerning a to e and u.

In some cases this information need not be more than what was stated above concerning their ranges (-2^r to 2^r for a to e, -2^p to 2^p for u), in others it may require additional information. But in any case the problem of foreseeing the sizes of the quantities which the calculation will produce (as intermediate or as final results), should be viewed as a mathematical problem in its own right. (Cf. the discussion of the "first stage" or preparing a problem, in 7.9.) This problem of evaluating sizes may be trivially simple, as in the present case; or moderately complex, as in most problems that require extensive calculations; or really difficult, as e.g. in various partial differential equation problems. It may even be, from the mathematical point of view, the crux of the whole problem, i.e. the actual problem may require more effective calculating than this subsidiary, evaluational problem, but the latter may require more mathematical insight than the former. This can happen for very important and practical problems, e.g. it is so for n simultaneous linear or non-linear equations in n variables for large values of n. (We will discuss this particular subject elsewhere.) To conclude, we note that the estimation of errors, and of the precision or significance that is obtainable in the result, is quite closely connected with the estimation of sizes, and that our above remarks apply to these, too. (All of these are covered by the "first stage" discussed in 7.9, as referred to above.)

Second: Under suitable conditions it may be possible to avoid the mathematical analysis that is required to control the sizes (as described above), by instructing the machine to rearrange the calculation, either continuously (i.e. for every number that is produced) or at suitable selected critical moments, so that all scales are appropriately readjusted and all numbers kept to the permissible size (i.e. between -1 and 1). The *continuous* readjusting is, of course, wasteful in time and memory capacity, while the *critical* (occasional) readjusting requires more or less mathematical insight. The latter is preferable, since it is almost always inadvisable to neglect a mathematical analysis of the problem under consideration.

Examples of how to code the instructions to secure critical size readjustments in various problems will be given subsequently.

Third: The process of continuous readjustment can be automatized and built into a machine, as a *floating binary point*. We referred to it already in 6.6.7 in the first part of this report. We formulated there various criticisms which caused us to exclude this facility, at any rate, from the first model of our machine. The above discussion reemphasizes this point: The floating binary point provides continuous size readjusting, while we prefer critical readjusting, and that only to the extent to which it is really needed, i.e. inasmuch as the sizes are not foreseeable without undue effort. Besides the floating binary point represents an effort to render a thorough mathematical understanding of at least a part of the problem unnecessary, and we feel that this is a step in a doubtful direction.

8.7 Let us now extend the Problems 1 and 2 in a different respect. This extension will bring in a simple induction, and thus the first complication of a logical nature.

Problem 3.

Same as Problem 1 with this change: The calculation is desired for I values of u: $u_1, \ldots, u_I$, giving these v's: $v_1, \ldots, v_I$, respectively. Each pair u_i, v_i ($i = 1, \ldots, I$) is stored or to be stored at two neighboring locations m + 2i-2, m + 2i-1, the whole covering the interval of locations from m to m + 2I-1. m and I are not given in advance, but stored at certain, given, memory locations. ---

This is the first one of our Problems to require a flow diagram. We will therefore apply the principles of Chapter 7, and more particularly of paragraph 7.9, to their full extent, and avoid any shortcuts that were possible in connection with our two previous problems.

Let A be the storage area corresponding to the interval of locations from m to m + 2I-1. In this way its positions will be A.1, ... 2I, where A.j corresponds to m + j-1 (j = 1, ..., 2I). These positions, however, are supposed to be part of some other routine, already in the machine, and therefore they need not appear when we pass (at the end of the whole procedure) to the final enumeration of the coded sequence.

Let B be the storage area which holds the given data (the constants) of the problem: a to e and m, I. m is a position mark, we will therefore store m_0 instead. I is really relevant in the combination m + 2I, which is a position mark (the storage area A is an interval that begins at m and ends just before m + 2I), we will therefore store $(m + 2I)_0$ instead of I. We store a to e at B.1, .., 5 and m_0, $(m + 2I)_0$ at B.6, 7. Finally, storage is required for the induction variable i. i is really relevant in the combinations m + 2i-2, m + 2i-1, which are position marks (for A.2i-1 containing u_i and for A.2i designated for v_i), we select m + 2i-2, and we will store $(m + 2i-2)_0$ instead. This will be stored at C.

These things being understood, we can proceed to drawing the flow diagram. We will use tabulated storage. A.2i-1 (i = 1, ..., I) and B.1, ...,7 are the fixed storage (the u_i and the a, ..., e, m, I, respectively), while A.2i (i = 1, .., I) and C are the variable storage (the v_i and i, respectively). The complete flow diagram is shown in Figure 8.1.

FIGURE 8.1

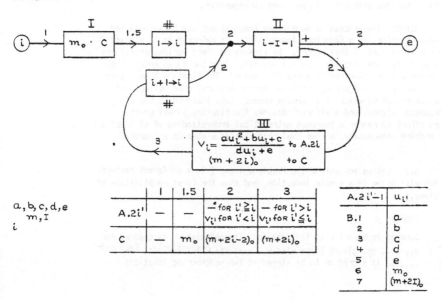

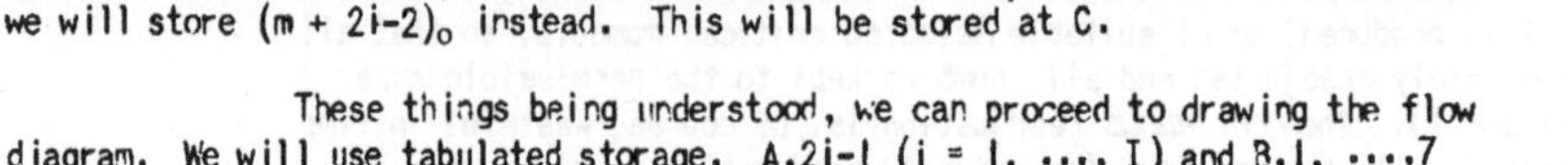

a, b, c, d, e
m, I
i

	1	1.5	2	3
A.2i'	—	—	— for i' ≧ i $v_{i'}$ for i' < i	— for i' > i $v_{i'}$ for i' ≦ i
C	—	m_0	$(m+2i-2)_0$	$(m+2i)_0$

A.2i'−1	$u_{i'}$
B.1	a
2	b
3	c
4	d
5	e
6	m_0
7	$(m+2I)_0$

We are left with the task of doing the static coding. This deals with the boxes I to III. We begin with the preliminary enumerations.

The coding of I is obvious:

I,1	B.6		Ac	n_0'
2	C		C	n_0'

The coding of II is also quite simple, we must only observe that $(m + 2i-2)_0$ - $(m + 2I)_0$ has the same sign as i-I-1. The code follows:

II,1	C		Ac	$(m + 2i-2)_0$
2	B.7	h-	Ac	$(m + 2i-2)_0$ $(m + 2I)_0$
3		Cc		

Here e is the position at which we want the control to finish the problem.

The coding of III is effected mainly after the pattern of Problem 2. This applies to it inasmuch as it produces v_i, the additional task of producing $(m + 2i)_0$ is trivial. There is, however, this to be noted: In these operations $(m + 2i-2)_0$ must be used to produce $(m + 2i-1)_0$, and $(m + 2i)_0$. Hence the number I_0 is needed. This number must be stored somewhere. This is not exactly local storage, because we are not dealing with a transient space requirement of III. It is fixed storage, and we assign to it the position B.8. Now the coding can be done as follows:

III,1	C		Ac	$(m + 2i-2)_0$
2	III,3	Sp	III,3	$(m + 2i-2)_0$
3	-			
[	m + 2i-2	]	Ac	u_i
4	s.I		s.I	u_i
5	C		Ac	$(m + 2i-2)_0$
6	B.8	h	Ac	$(m + 2i-1)_0$
7	III,22	Sp	III,22	m + 2i-1 S
8	B.8	h	Ac	$(m + 2i)_0$
9	C	S	C	$(m + 2i)_0$
10	B.4	R	R	d
11	s.I	x	Ac	du_i
12	B.5	h	Ac	$du_i + e$
13	s.2	S	s.2	$du_i + e$
14	B.I	R	R	a
15	s.I	x	Ac	au_i
16	B.2	h	Ac	$au_i + b$
17	s.3	S	s.3	$au_i + b$
18	s.3	R	R	$au_i + b$
19	s.I	x	Ac	$au_i^2 + bu_i$
20	B.3	h	Ac	$au_i^2 + bu_i + c$
21	s.2		R	$v_i = \dfrac{au_i^2 + bu_i + c}{du_i + e}$

III,22	—	S			
[	m + 2i-1	S]		A.2i	v_i'
III,23	II,1	C			

We had to use some local storage: s.1-3.

        We now pass to the final enumeration, assigning B.1-8, C and s.1-3 their actual values, pairing the 27 orders |1,1, 2, |II,1-3, |III,1-22 to 14 words, and then assigning |1,1, ..., |III,22 their actual values. Let these 14 words be 0 to 13, then B.1-8 may be 14 to 21, C may be 22, |III|, s.1-3 may be 23-25, and we can put e equal to, say 26. So we obtain this coded sequence:

0	19 , 22 S		8	18 h , 24 S		17	d	
1	22 , 20 h-		9	14 R , 23 x		18	e	
2	26 Cc, 22		10	15 h , 25 S		19	m_0	
3	3 Sp', —		11	25 R , 23 x		20	$(m + 2I)_0$	
			12	16 h , 24 ÷		21	I_0	
4	23 S , 22		13	— S , 1 C		22	—	
5	21 h , 13 Sp		14	a		23	—	
6	21 h , 22 S		15	b		24	—	
7	17 R , 23 x		16	c		25	—	

The durations may be estimated as follows:

I: 75 μ, II: (2 × 75 - 25) μ = 125 μ.
III: (12 × 75 - 25 + 3 × 70 + 120) μ = 1,205 μ.
Total: I + II + (II + III) × I = (75 + 125 + (125 + 1,205) I) μ =
   = (1330 I + 200) μ ≈ 1.3 I m.

8.8 We give now an example of how to code instructions for a critical readjustment, as mentioned in the second remark in 8.6.

Problem 4.

        The variables u, v are stored in certain, given locations. It is desired to form $\frac{u}{v}$ and to determine its size, in the following sense: Find the integer n (= 0, ± 1, ± 2, ...) for which $2^{n-1} \leq |\frac{u}{v}| < 2^n$, and then form $w = 2^{-n} \frac{u}{v}$. This scheme will fail if either u or v is zero. sense whether this is the case. —

        We could also require that other things, e.g. the number of significant digits in u and v, be determined. But we will not introduce these additional complications here, they can be handled along the same lines as the main Problem.

        In our machine u, v ≠ 0 imply $2^{-39} \leq |u|, |v| < 1$, hence $2^{-39} < |\frac{u}{v}| < 2^{+39}$, hence $-38 \leq n \leq 39$. We will therefore signalize v = 0 with n = 50, and u = 0 (and v ≠ 0) with n = -50. We will begin the calculation by sensing whether v = 0 and whether u = 0. This is decided by testing $-|v| \geq 0$ and $-|u| \geq 0$. If neither is the case, we decide whether n is going to be ≤ 0 or > 0. This is done by testing whether $|u| - |v| < 0$, or not. After this we determine n by two obvious, alternative inductions.

Let u, v be stored at A.1,2. Since n has the values -38, -37, ..., 39 and -50, 50, we store $2^{-39}n$ instead. We store n and $2^{-n}\frac{u}{v}$, when they are formed, at B.1,2. For u = 0 or v = 0, i.e. when n = -50 or 50, we will not form $\frac{u}{v}$, or rather $2^{-n}\frac{u}{v}$. In this case, then, the content of B.2 is irrelevant. In any case, we denote the content of B.2 by t.

In searching for n we will need an induction variable i, which will move over 0, -1, -2, ... to n if we have $n \leq 0$, and over 1, 2, ... to n if we have $n > 0$. It is convenient to treat i the same way as n: Store $2^{-39}i$ instead of i. Since we propose to use distributed storage, we need not enumerate further storage locations, we are free to introduce them as the need for them arises, while we are drawing the flow diagram.

These indications should suffice to clarify the evolution and the significance of all parts of the flow diagram shown in Figure 8.2.

The boxes I-XIV require static coding. Before we do that, we make three remarks, which embody principles that will apply correspondingly to our future static coding operations as well. They are the following:

First: Some additional fixed storage will be required. These are certain numbers to which specific reference will have to be made, specifically $2^{-39}\cdot 50$, 2^{-39}, 0. We will store them at the positions D.1, 2, 3, respectively. We will state these explictly at the points where they become necessary, by interrupting the coding column there and stating the storage.

Second: Every box coding either ends with an unconditional transfer order (type xC, xC') or else it involves the assumption that the control goes on from there to a definite position. (Ending with a conditional transfer order -- type xCc, xCc' -- presents an intermediate case: We have the first one or the second one of the two above alternatives according to whether the transfer condition is or is not fulfilled.) If this latter alternarive is present, then this position (i.e. its Roman box numeral and its number within its box) has to be named at the end of the box coding , with a reference "to ...".

Third: It will be found convenient to introduce an extra operation box, in addition to those mentioned in the flow diagram. The reason is, that two operations which should be carried out in two operation boxes immediately preceding a confluence, can be merged and carried out after the confluence. (The two boxes in question are XIII, XIV, the new box is XV.)

We now proceed to carry out the static coding of all boxes in one, continuous operation:

I,1	A.2	-M	Ac	$-\lvert v\rvert$
2	II,1	Cc		
(to III,1)				
D.1	$2^{-39}\cdot 50$			
II,1	D.1		Ac	$2^{-39}\cdot 50$
2	B.1	S	B.1	$2^{-39}\cdot 50$
3	e	C		
III,1	A.1	-M	Ac	$-\lvert u\rvert$
2	IV,1	Cc		
(to V.1)				

FIGURE 8.2

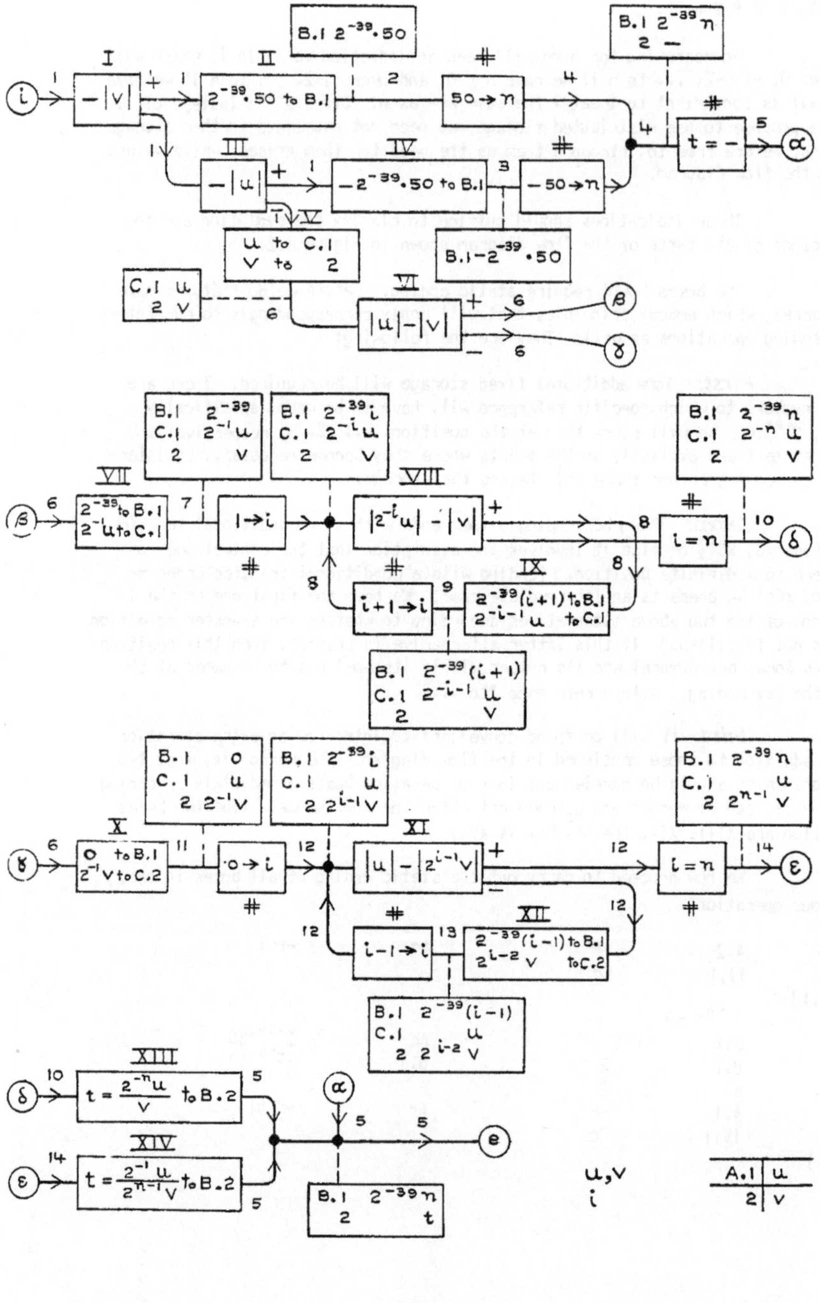
I
−|v|
II
2-39·50 to B.1
B.1 2-39·50
50 → n
B.1 2-39 n
2 —
t = −
III
−|u|
IV
−2-39·50 to B.1
−50 → n
B.1 −2-39·50
V
u to C.1
v to 2
C.1 u
2 v
VI
|u| − |v|
VII
2-39 to B.1
2-1u to C.1
1 → i
VIII
|2-iu| − |v|
i = n
IX
2-39(i+1) to B.1
2-i-1u to C.1
i+1 → i
X
0 to B.1
2-1v to C.2
0 → i
XI
|u| − |2i-1v|
XII
2-39(i−1) to B.1
2i-2v to C.2
i−1 → i
XIII
t = 2-nu / v to B.2
XIV
t = 2-1u / 2n-1v to B.2
B.1 2-39n
2 t
u, v
i
A.1 u
2 v

IV,1	D.1	-	Ac	$-2^{-39}\cdot 50$
2	B.1	S	B.1	$-2^{-39}\cdot 50$
3	e	C		
V,1	A.1		Ac	u
2	C.1		C.1	u
3	A.2		Ac	v
4	C.2	S	C.2	
(to VI,1)				
VI,1	A.1	M	Ac	$\|u\|$
2	A.2	h-M	Ac	$\|u\| - \|v\|$
3	VII,1	Cc		
(to X,1)				
D.2	2^{-39}			
VII,1	D.2		Ac	2^{-39}
2	B.1		B.1	2^{-39}
3	C.1		Ac	u
4		R	Ac	$2^{-1}u$
5	C.1	S	C.1	$2^{-1}u$
(to VIII,1)				
VIII,1	C.1	M	Ac	$\|2^{-i}u\|$
2	C.2	h-M	Ac	$\|2^{-i}u\| - \|v\|$
3	IX,1	Cc		
(to XIII,1)				
IX,1	B.1		Ac	$2^{-39}i$
2	D.2	h	Ac	$2^{-39}(i+1)$
3	B.1	S	B.1	$2^{-39}(i+1)$
4	C.1		Ac	$2^{-i}u$
5		R	Ac	$2^{-i-1}u$
6	C.1	S	C.1	$2^{-i-1}u$
7	VIII,1	C		
D.3	0			
X,1	D.3		Ac	0
2	B.1		B.1	0
3	C.2		Ac	v
4		R	Ac	$2^{-1}v$
5	C.2	S	C.2	$2^{-1}v$
(to XI,1)				
XI,1	C.1	M	Ac	$\|u\|$
2	C.2	h-M	Ac	$\|u\| - \|2^{i-1}v\|$
3	XIV,1	Cc		
(to XII,1)				
XII,1	B.1		Ac	$2^{-39}i$
2	D.2	-h	Ac	$2^{-39}(i-1)$
3	B.1	S	B.1	$2^{-39}(i-1)$
4	C.2		Ac	$2^{i-1}v$
5		R	Ac	$2^{i-2}v$
6	C.2	S	C.2	$2^{i-2}v$
7	XI,1	C		
XIII,1	C.1		Ac	$2^{-n}u$
2	C.2		R	$t = \frac{2^{-n}u}{v}$
(to XV,1)				

2	B.2	S	B.2
3	e	C	

We now proceed to the final enumeration. We must order the boxes I-XV for this purpose in such a manner, that the indications "to ..." are obeyed. Any sequence which is made necessary by an indication "to ... " will be marked by a comma, and every sequence which is based on an arbitrary choice is marked by a semicolon. In this manner we may set the following ordering: I, III, V, VI, X, XI, XII; II; IV; VII, VIII, XIII; IX; XIV, XV. The ordering is imperfect, inasmuch as XV should have been the immediate successor of XIII as well as of XIV, since only one was possible, we chose XIV. Hence it is necessary to add at the end of XIII an unconditional transfer order to XV:

XIII,8	XV,1	C

It remains for us to assign A.1,2, B.1,2, C.1,2, D.1-3 their actual values, to pair the 56 orders I,1,2, II,1-3, III,1,2, IV,1-3, V,1-4, VI,1-3, VII,1-5, VIII,1-3, IX,1-7, X,1-5, XI,1-3, XII,1-7, XIII,1-3, XIV,1-3, XV,1-3, to 28 words, and then to assign 1,1,..., XV,3 their actual values. These assignments are sufficiently numerous to justify making a small table:

I,1-2	0, 0'	XII,1-7	9'- 12'	IX,1-7	21'-24'
III,1-2	1, 1'	II,1-3	13- 14	XIV,1-8	25 -26
V,1-4	2 -3'	IV,1-3	14'-15'	XV,1-3	26'-27'
VI,1-3	4 -5	VII,1-5	16 -18	A.1-2	28, 29
X,1-5	5'-7'	VIII,1-3	18'-19'	B.1-2	30, 31
XI,1-3	8 -9	XIII,1-3	20 -21	C.1-2	32, 33
				D.1-3	34 -36

Now we obtain this coded sequence:

0	29 -M,	13 Cc	12	33 S,	8 C	24	32 S,	18 C'
1	28 -M,	14 Cc'	13	34 ,	30 S	25	32 ,	R
2	28 ,	32 S	14	e C,	34 -	26	33 ÷,	A
3	29 ,	33 S	15	30 S,	e C	27	31 S,	e C
4	28 M,	29 h-M	16	35 ,	30 S	28	u	
5	16 Cc,	36	17	32 ,	R	29	v	
6	30 S,	33	18	32 S,	32 M	30	-	
7	R,	33 S	19	33 h-M,	21 Cc'	31	-	
8	32 M,	33 h-M	20	32 ,	33 ÷	32	-	
9	25 Cc,	30	21	26 C',	30	33	-	
10	35 -h,	30 S	22	35 h,	30 S	34	$2^{-39}\cdot 50$	
11	33 ,	R	23	32 ,	R	35	2^{-39}	
						36	0	

In estimating durations, we have to pay attention to some points which did not come up in Problems 1-3, namely: The flow diagram of Figure 8.2 is never traversed by the control C in its entirety. The course of C is, instead, dependent on the behavior of u, v and n. Specifically, its course is as follows:

I, then
for v = 0: II (end),
for v ≠ 0: III, then
for u = 0: IV (end),
for u ≠ 0: V, VI, then
for n > 0: VII, VIII x n, IX x (n-1), XIII } , then
for n ≤ 0: X, XI x (-n + 1), XII x (-n), XIV }
XV.

The durations of the individual boxes are:

I : 75 μ	II : 125 μ	III : 75 μ	IV: 125 μ	V : 150 μ
VI: 125 μ	VII : 180 μ	VIII : 125 μ	IX: 255 μ	X : 180 μ
XI: 125 μ	XII : 255 μ	XIII : 245 μ	XIV: 225 μ	XV : 105 μ

Hence we have these total durations:

For v = 0 : 200 μ = .2 m,
for v ≠ 0, u = 0 : 275 μ ≈ .3 m,
for v ≠ 0, u ≠ 0, n > 0: (380 (n - 1) + 1080) μ ≈ (.4 (n - 1) + 1) m,
for v ≠ 0, u ≠ 0, n ≤ 0: (380 (-n) + 1060) μ ≈ (.4 (-n) + 1) m.

8.9 It seems appropriate to make at this point a remark on how to correct errors that are discovered at any stage of coding, including the possibility that they are discovered after the coding has been completed. Together with correcting actual errors, one should also consider the case where the coding is proceeding, or has been concluded, without fault, when (for any reason whatever) a decision is made to apply certain changes. Changes of both types (i.e. either corrections of actual errors or voluntary transitions from one correct code to another one) fall clearly into three classes: Omissions of certain parts of the coded sequence; insertions at a certain places into a coded sequence; replacement of a certain part of the coded sequence by another coded sequence (the new having the same length as the old, or a smaller length, or a greater length). The important thing is, of course, that it should be possible to effect such modifications, without losing the result of any of the work done on those parts of the coded sequence which are not to be modified -- i.e. without any necessity to "start all over again" in whole or in part.

There is, of course, no problem if the coding has not yet progressed very far. The dynamic stage, i.e. the flow diagram, is easy to correct: Omissions can be effected by erasing or crossing out the undesirable region, followed by "shorting", i.e. taking the track directly from its immediate predecessors to its immediate successors. This "shorting" may also be achieved by introducing (fixed) remote connections. Insertions can be effected by drawing the new region separately, and then detouring the track to it and back from it. Again fixed remote connections may be helpful. Replacements can be effected by combining these two techniques -- provided that simple erasing and redrawing of the affected region is not practical.

Even in the static coding there is no real problem, as long as the stage of the preliminary enumeration has not been passed. Erasing or crossing out, and re-writing at the same place or elsewhere, will take care of everything considering the great flexibility of our methods of (preliminary) enumeration. (Cf. Note I in 7.4 and the discussion of the preliminary enumeration in 7.9.)

The difficulties arise only when the final enumeration has already been achieved, i.e. when the coded sequence has already been formulated in its final form, or when it has even been put on the magnetic wire or tape. In this case, any omission, or insertion, or replacement of any piece by one of different length, may force the rewriting of considerable portions of the coded sequence, which may lie in any part of it. Indeed, it will change primarily the physical positions of the words and their (final) numbers, inasmuch as they lie after the region of the changes. If the change in length amounts to an odd number of half words (orders), it will even change the left or right position of all subsequent half words (orders) within their words. Secondarily, it will change the x in all S(x) everywhere in the coded sequence, which refer to words or half words whose position was changed primarily.

Let us discuss how changes are effected on a coded sequence that has been formulated (on paper) in its final enumeration. We will afterwards touch quite briefly upon the treatment of changes in a coded sequence which has already reached the magnetic wire or tape stage. This latter phase can only receive a preliminary consideration in this report, since its exact treatment is affected by a number of engineering details which cannot be taken up here. It will become clear, however, that the treatment of this problem is in any event possible along essentially the same lines as the treatment of the first mentioned problem: That one of a coded sequence formulated (on paper) in its final enumeration.

The best way to deal with this problem is by means of a specific example. We will use the final coded sequence solving Problem 4, as given near the end of 8.8, for this purpose. Our changes will consist of corrections of actual errors (we will introduce such errors into that sequence), changes which correspond to voluntary transitions from one correct code to another one offer clearly no new problems.

We consider therefore the final coded sequence 0-36 in 8.8. We take up several hypothetical errors and their corrections:

First: Assume that by some mistake three orders which do not belong in this sequence, had been placed into the three half word positions beginning with the first half of 14. They occupy 14-15, and our actual orders 14-27' are at 15'-29, and the storage words 28-36 are at 30-38. All references, on the other hand, are correct with respect to this positioning. It is desired to remove the erroneous orders, 14-15 (this is the old notation).

It suffices to replace the order 14 by an unconditional transfer

14 15 C' ,

14', 15 may be left standing. After 13' C gets to 14, and does nothing there, but is immediately transferred to 15', where it belongs.

Second: Assume that by some mistake the two orders 6', 7 are missing. The actual orders 7'-27' are at 6'-26', and the storage words 28-36 are at 27-35. All references, on the other hand, are correct with respect to this positioning.

It is desired to insert the missing orders 6', 7 (this is the old notation).

Choose a location in the memory, where an unused interval of sufficient length begins. This may be the first position following the coded sequence in question, or any other position. We assume the former, and choose accordingly 36 (this is the new notation). We displace the order immediately preceding the missing ones, in the present case 6, and replace it by an unconditional transfer

6	36	C ,		

Beginning with 36, we add the removed order 6 as well as the missing orders 6', 7 , and conclude with an unconditional transfer back to the order immediately following upon the missing ones. This gives

36	29	S ,	32	
37		R ,	6	C'

(The 32 in order 36 is the new notation for the old 33.) If the sequence contained any references to 6 (which does not happen to be the case here), they have to be replaced by references to 36.

Third: Assume that certain orders in the coded sequence were erroneously replaced by other orders. The number of erroneous orders may or may not be equal to the number of the correct orders which they displace. In the latter case the entire subsequent part of the coded sequence is displaced, but we assume that at any rate all references are correct with respect to the actual positioning.

If the correct sequence has the same length as the erroneous one, we can simply insert it in the place of the latter. If the correct sequence is shorter, then we insert it, and proceed from there on as in the first case. If the correct sequence is longer, then we insert as much of it as possible, and proceed from there on as in the second case. --

It will be clear from the above, that changing a coded sequence which is already on the magnetic wire or tape can be effected according to the same principles. This requires, of course, a device which can change single words on the magnetic wire or tape. We expect, that the main machine itself, as well as the special typewriter which produces the magnetic wire or tape under manual control, will be able to do this. This means, that the desired changes can be effected either by the main machine, with appropriate instructions, or by the typewriter, manually. The latter procedure will probably be preferred in most cases. (Cf. however Chapter 12 , for a somewhat related situation where the first procedure turns out to be preferable.)

8.10 Problems 1-4 were of a somewhat artificial character, designed to illustrate certain salient points of our method of coding, but only indirectly and very incompletely typifying actual mathematical problems. From now on we will lay a greater stress on the realism of our examples. We will still select them so that they illustrate various important aspects of our coding method, but we will also consider it important that they should represent actual and relevant mathematical problems. Furthermore, we will proceed from typical parts, which occur usually as parts of larger problems, to increasingly completely formulated self-contained problems.

We begin with a code for the operation of square rooting. This is an important problem, because our machine is not expected to contain a square rooting organ, i.e. square rooting is omitted from its list of basic arithmetical operations.

Problem 5.

The variable u is stored in a given memory location. It is desired to form $v = \sqrt{u}$ and to store it in another, given, memory location.

v should be obtained by iterating the operation $z \rightarrow \frac{1}{2}(z + \frac{u}{z})$. --- This iterative process can be written as follows:

$$z_{i+1} = \frac{1}{2}(z_i + \frac{u}{z_i}),\ i = 0, 1, 2, \ldots,\ \lim_{i \to \infty} z_i = v = \sqrt{u}\ .$$

Two questions remain: First, what should be the value of the initial z_0? Second, for what $i = i_0$ should the iteration stop, i.e. which z_{i_0} may be identified with $\sqrt{u}$? The answers are easy:

Answer to the first question: Since we prefer to avoid adjustments of size, it is desirable that all numbers with which we deal lie between -1 and 1. u will be between -1 and 1, and it is necessary to have $u \geq 0$, hence $0 \leq u < 1$. z_i must be between -1 and 1, and it is desirable to have $z_i \geq 0$, hence $0 \leq z_i < 1$. Consequently the only numbers, whose sizes must be watched are $\frac{u}{z_i}$ and $z_i + \frac{u}{z_i}$. The first requires $u < z_i$, i.e. $u < z_i < 1$. The second one may be circumvented, by forming the desired $\frac{1}{2}(z_i + \frac{u}{z_i})$ in this way: $\frac{1}{2}(z_i + \frac{u}{z_i}) = \frac{1}{2}(\frac{u}{z_i} - z_i) + z_i$. Next clearly

$$(*) \qquad z_{i+1} - \sqrt{u} = \frac{1}{2z_i}(z_i - \sqrt{u})^2\ .$$

Hence $z_{i+1} \geq \sqrt{u}$, i.e. all $z_i \geq \sqrt{u}$ with the possible exception of z_0. Choose therefore $z_0 \geq \sqrt{u}$, i.e. $\sqrt{u} \leq z_0 \leq 1$. Then all $z_i \geq \sqrt{u}$, and so $z_i > u$ ($u < 1$ implies $u < \sqrt{u}$) as desired. Now $z_i \geq \sqrt{u}$ gives $z_i \geq \frac{u}{z_i}$, hence $z_i \geq z_{i+1}$, and so

$$1 \geq z_0 \geq z_1 \geq \ldots \geq \text{ and converging to } \sqrt{u}\ .$$

In choosing z_0 we may choose any number that is reliably $\geq \sqrt{u}$. The simplest choice is $z_0 = 1$. This gives $z_1 = \frac{1}{2}(1+u)$. Since this expression involves no division, we might as well begin with $i = 1$ and $z_1 = \frac{1}{2}(1 + u)$.

Answer to the second question: u is known to within an error $\approx \frac{1}{2} \cdot 2^{-39} = 2^{-40}$, hence $\sqrt{u}$ is known within $\frac{d\sqrt{u}}{du} = \frac{1}{2\sqrt{u}}$ times this error, i.e.

$$\varepsilon_u = 2^{-41}\sqrt{u}^{\,-1}\ .$$

For $u \neq 0$, as u varies from 1 to 2^{-39}, ε_u varies from 2^{-41} to $2^{-21.5} \approx .7 \times 2^{-21}$. For $u = 0$ the above formula breaks down, however the error is clearly $\sqrt{2^{-40}} - \sqrt{0} = 2^{-20}$, i.e. $\varepsilon_0 = 2^{-20}$. We assume $u \neq 0$, since the case $u = 0$ is quite harmless. We should stop when z_i lies within ε_u of $\sqrt{u}$. Assume that the last step is the one from z_{i_0} to z_{i_0+1}, then we want $|z_{i_0+1} - \sqrt{u}| \lessapprox \varepsilon_u$. This means $|z_{i_0} - \sqrt{u}|^2 = 2z_{i_0} | z_{i_0+1} - \sqrt{u}| \lessapprox 2^{-40}$, $| z_{i_0} - \sqrt{u}| \lessapprox 2^{-20}$ or, since $z_{i_0+1} - \sqrt{u}$ is negligible compared to $z_{i_0} - \sqrt{u}$, $|z_{i_0} - z_{i_0+1}| \lessapprox 2^{-20}$. Using $z_{i_0} > z_{i_0+1}$, and making a safe overestimate gives

$$(**) \qquad z_{i_0} - z_{i_0+1} \leq 2^{-19}.$$

Using (*), it is easily seen, that (**), or the equivalent

$$(\overset{*}{**}) \qquad z_{i_0} - \sqrt{u} \lessapprox 2^{-19},$$

obtains for $i_0 = 1$ if $1-2^{-8} \leq u < 1$, for $i_0 = 2$ if $1-2^{-8} \leq u \leq 1-2^{-8}$, for $i_0 = 3$ if $2^{-1} \leq u \leq 1-2^{-2}$, for $i_0 = 4$ if $5^{-1} \leq u \leq 2^{-1}$, ..., $i_0 = 19$ for $u = 2^{-40}$ or 0. Hence for most values of u, specifically for $.2 \leq u \leq .875$ we need 3-4 iterations. Larger numbers i_0 of iterations are needed when u is small, if u is of the order 2^{-2j}, then i_0 lies between $j-3$ and $j-1$. It is therefore debatable, whether we should not adjust u first, by bringing it into the "favorable" range $u \geq .2$ by iterated multiplications with 2^2. In order to simplify matters, we will not do this, and operate with u unadjusted, relying on (**) only. The disadvantage thus incurred is certainly not serious.

This arrangement of the iterative process has the effect, that there is no need to store the iteration index i (but we will, of course, have to refer to i in drawing the flow diagram). We store u at A, and $v = \sqrt{u}$, when it is formed, at B. We use distributed storage. In the preceding problems we refrained from using the storage location, which is reserved for the final result, for changing storage, too. We will now do this.

The flow diagram should now present no difficulties, it is shown in Figure 8.3

FIGURE 8.3

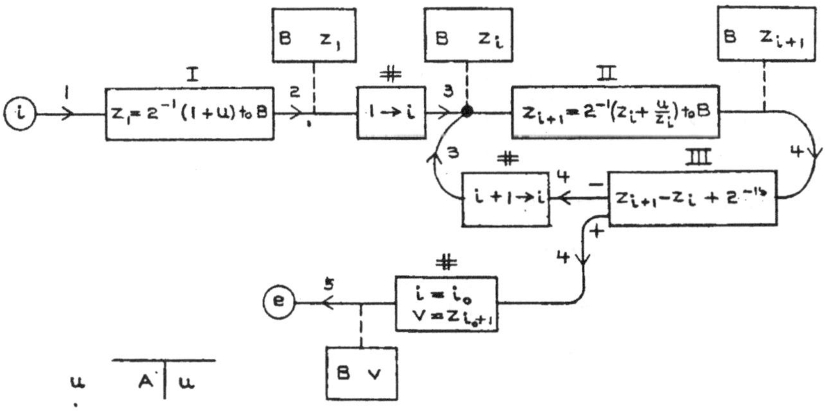

The boxes I-III require static coding. We proceed in the same sense as we did in coding Problem 4, and we code all boxes in one, continuous operation:

I,1	A		Ac	u
2		R	Ac	$2^{-1}u$
C.1	2^{-1}			
I,3	C.1	h	Ac	$z_1 = 2^{-1}(1+u)$
4	B	S	B	z_1
(to II,1)				
II,1	A		Ac	u
2	B	÷	R	$\frac{u}{z_i}$
3		A	Ac	$\frac{u}{z_i}$
4	B	h-	Ac	$\frac{u}{z_i} - z_i$
5		R	Ac	$z_{i+1} - z_i = 2^{-1}\left(\frac{u}{z_i} - z_i\right)$
C.2	—			
6	C.2	S	C.2	$z_{i+1} - z_i$
7	B	h	Ac	z_{i+1}
8	B	S	B	z_{i+1}
(to III,1)				
III,1	C.2		Ac	$z_{i+1} - z_i$
C.3	2^{-19}			
III,2	C.3	h	Ac	$z_{i+1} - z_i + 2^{-19}$
3	e	Cc		
(to II,1)				

Next, we order the boxes, in the same sense as we did in coding Problem 4. The ordering I, II, III results with one imperfection: II should have been the immediate successor of III. This necessitates an extra order

III,4	II,1	C

To conclude, we have to assign A, B, C.1-3 their actual values. pair the 16 orders I,1-4, II,1-8, III,1-4 to 8 words, and then assign I,1,...,III,4 their actual values. These are expressed in this table:

I,1-4	0 - 1'	III,1-4	6 - 7'	B	9
II,1-8	2 - 5'	A	8	C.1-3	10 - 12

Now we obtain this coded sequence:

0	8 , R	4	R, 11 S	8	u
1	10 h, 9 S	5	9 h, 9 S	9	-
2	8 , 9 ÷	6	11 , 12 h	10	2^{-1}
3	A, 9 h-	7	e Cc, .2 C	11	-
				12	2^{-19}

The durations may be estimated as follows:

I: $130\,\mu$, II: $380\,\mu$, III: $150\,\mu$.

Total: $\text{I} + (\text{II} + \text{III}) \times i_o = (130 + (380 + 150)\, i_o)\,\mu =$
$= (530\, i_o + 130)\,\mu \approx (.6\, i_o + .1)$ m.

As we saw above, a reasonable mean value for i is 4. This gives a duration of 2.5 m.

9.0 CODING OF PROBLEMS DEALING WITH THE DIGITAL CHARACTER OF THE NUMBERS PROCESSED BY THE MACHINE.

9.1 We will code in this chapter several problems, which fall into two classes.

First: The conversion of numbers given in the decimal system into their binary form, and conversely.

Second: The instructions which are necessary to make the machine work with numbers of more than 40 binary digits.

These problems are of considerable interest for a variety of reasons, and it seems worth while to dwell on these somewhat extensively.

We note, to begin with, that both classes of problems furnish examples of problems and procedures that deviate from the normal schemes of algebra, and hence differ very relevantly from the problems that we have handled so far. Indeed, up to now (as well as in all cases that we will consider after this chapter) the 40 digit aggregate, which is the number contained in a word, was the standard unit which we manipulated. We applied to such units the elementary operations of arithmetics (addition, subtraction, multiplication, division), but we never combined them in any way except through these operations, and we never broke any such unit up into smaller parts. In the problems of this chapter we will have to break up numbers into their digits, operate on these digits (or rather on certain digital aggregates) individually, combine numbers in new ways, etc.; that is, our operations will be in many important phases logical and combinatorial rather than algebraically mathematical.

Next, we have to point out the practical importance of the two classes of problems to which we have referred. The second class can be disposed of by a simpler discussion than the first one; therefore we will consider the second class first.

9.2 Our standard numbers have 40 binary digits, or rather a sign and 39 binary digits; hence they represent a relative precision of $2^{-39} = 1.8 \cdot 10^{-12}$. Comparing this to the length of the interval which is available for them (from -1 to 1; i.e., length 2), the precision becomes $2^{-40} = .9 \cdot 10^{-12}$. At any rate, we have here a precision of about 12 decimals. This is a considerable precision, and adequate or more than adequate for most problems that are of interest at this moment. Yet there are even now certain exceptions; e.g., the summation of certain series, the evaluation of certain expressions by recursion formulae, the integration of some non-linear (total) differential equations, etc. Furthermore, these cases, where 12 decimals (or their equivalents) are inadequate, are likely to become more frequent as more involved computing problems are tackled in the future. It will therefore become increasingly difficult to choose an "appropriate" number of digits for an all-purpose machine, especially since an unusually high number of of digits (for the standard number) would render it very inefficient when it deals with problems that require ordinary precisions (and these will probably continue to constitute the majority of all applications). This inefficiency would extend to the use of the available memory capacity, the amount of equipment tied up in the parallel digital channels of the machine, and the time required to perform multiplications and divisions.

Consequently, it is important that the machine should allow being operated at varying numbers of digits; i.e. at varying levels of digital precision. One might consider building in alternative facilities for alternative precision levels, controlled by manual switches, or partly or wholly automatically. We prefer, however, to achieve this without any physical changes or extra organs of the 40 binary digit machine. This can be done purely logically; that is, it is possible to instruct the machine to treat k words; i.e., k aggregates of a sign and 39 binary digits, as if they were forming a number with a sign of 39 k binary digits. (k-l signs are wasted in this process.) This increases the precision to 2^{-39k}; that is, to $10^{-11.7k}$; that is, to any desired level. It does require, however, new instructions for the operations of arithmetics (addition, subtraction, multiplication, division), and it cannot fail to reduce the speed and the memory capacity of the machine. Nevertheless, it is perfectly reasonable as an emergency measure, when an exceptional problem requires unusual precision.

These new arithmetical instructions are provided by the problems of our second class.

9.3 Our machine will operate in the binary system. This is most satisfactory from the point of view of its arithmetical and its logical structure, but it creates certain problems in connection with its practical use.

In most cases the input data are given in the decimal system, and the output figures (the results) are also wanted in that system. It is therefore important to be able to effect the decimal-binary and the binary-decimal conversions by fast, automatic devices. On the other hand, the use of these devices should be optional; i.e., the machine should be able to receive and to emit data in the binary system, too. The reasons for this requirement deserve being stated explicitly.

The use of the decimal system is only justifiable for inputs (data) which originate (directly) in human operations; i.e., where the magnetic wire or tape that carries them has been produced manually (by typing): and for outputs (results) which are intended for (direct) human sensing; i.e., where the contents of the magnetic wire or tape in question are to be made readable in the ordinary sense (by printing). A good deal of the output of the machine is, however, destined solely for use as subsequent input of the machine -- in the same problem or in some later problem. For this material no conversion is needed, since it need never be moved out of the binary system. Thus there are, as a matter of principle, two categories within the information stored on the magnetic wires or tapes: First, information that must pass through human intellects (at its origin or at its final disposition); second, information that never reaches (directly) a human intellect and remains at all times strictly within the machine cycle. Only the first category need be transferable to and from the decimal notation; i.e., only it requires the conversions that we are going to consider. This circumstance is of great importance in dealing with these conversions, because it sets for them much less exigent standards of speed than would be desirable otherwise.

Indeed, information of the second category may be produced and consumed at a rate which is only limited by the speed of writing on or reading from the magnetic wire or tape. This can probably be done at rates of 50-100 thousand digits per second. We may, at least at first, use a more conservative arrangement of about 25 thousand digits per second. Even this means, for a 40 digit number, and with reasonable time allowances for certain checking features, etc., about 2 m per number. In order that the conversions do not slow down these processes appreciably, they would have to consume definitely less than 1 m each. Information of the first category, on the other hand, should not be produced or consumed faster than the rate corresponding to that of printing, typing, and what is more restrictive, human "understanding". Our 40 binary digit numbers correspond to numbers with about 10 decimal digits. It is difficult to assign a definite printing time to this, since one may print many digits in parallel. Thus, the standard IBM printer will print 80 decimal digits in somewhat less than a second. Various unconventional processes may be a good deal faster. On the other hand, typing will hardly ever be faster than 10 decimal digits per second, and any form of human "understanding" (which should be taken to mean at least critical reading) will be still slower. Even the making of tables for publication need not exceed this rate, and besides it is likely that the making of voluminous tables in book form will become less important with the introduction of very high speed machine computing. To the extent to which the machine is making "tables" for its own use, they fall into the second category; i.e. they should remain on the wire or tape medium, and never be converted into the decimal system. Hence, information of the first category can be handled satisfactorily with any conversion methods that are considerably shorter than a second. An upper limit of 100 m would seem more than safe.

These considerations show that our subdivision of the input and output information of the machine into the two categories discussed above is indeed relevant, and that it is also important to realize that the desirability of the conversion applies only to the first category. These things being understood, any conversion times up to 100 m turn out to be adequate.

The machine will easily effect the conversions in question at much faster rates than this. With small changes in its controls we could even keep within the 1 m, which would be desirable if all information were to be converted, but we have seen that this is altogether unnecessary. We will therefore make no changes at all. The conversion times arrived at in this manner will be of the order of 5 m.

In order to discuss these, one more preparation is necessary: If the machine is to effect binary-decimal and decimal-binary conversions, then we must agree on some method by which it can store decimal digits; i.e., express them in a binary notation. This must also be correlated with various characteristics of the wire or tape medium: How the decimal information is stored on it, how it can be put on it by typing, and extracted from it by printing.

We will now proceed to consider these matters.

9.4 In discussing typing and printing, one should keep in mind that, while it is necessary to be able to perform these operations in the decimal system, it is quite desirable to have the alternative possibility of doing them in the binary system. Since the inner functioning of the machine is binary, its extensive use will successively lead to an increasing familiarity with the binary system, at least as far as those are concerned who are directly engaged in operating it; making arithmetical and other estimates for it; sensing, diagnosing and correcting its malfunctions; etc. It is likely that the machine and the practice and theory of its functioning will produce genuinely binary problems. It is therefore probable that it will gradually cause, even in the interests of its users, a certain amount of a shift from decimal to binary procedures. The problem of typing, printing and (human) reading should therefore be faced in the binary as well as in the decimal system.

Consider, first, the binary case. The machine, as well as the wire or tape, treats a number as an aggregate of 40 binary digits, i.e., digits 0 or 1. To type in this fashion, and even more to print and to have to read in this fashion, would be definitely inefficient, since we are already conditioned to recongize and use about 80 different alphabetic, numerical and punctuation symbols. A string of 40 digits is definitely harder to remember and to manipulate than a string of 10 symbols, even if the former are only 0's and 1's, while the latter may be chosen freely from 10 decimal digits 0, ..., 9 or from 26 alphabetic symbols a, ..., z.

There is, however, an obvious way to circumvent this inconvenience. It consists of choosing a suitable k = 1, 2, 3, ..., and of typing, printing and reading not in terms of individual binary digits, but of groups of k such digits. This amounts, of course, to using effectively a base 2^k system instead of the base 2 (binary) system, but since the conversions between these two are trivial to the point of insignificance, and since the machine is binary, this seems a natural thing to do. Considering our reading habits and conditioning, 2^k should be of the order of 10 to 80; since it is to replace the number 10 of decimal digits, it is preferable to make it of the order of 10. Hence, k = 3 or k = 4 seem the logical choices.

For k = 3, 2^k = 8, we can denote these triads of binary digits by the decimal digits 0, ..., 7. It is best to let every decimal digit 0, ...,7 correspond to its binary expression (as an integer). This is an advantage. On the other hand, a printer that is organized in this fashion would be unable to print the decimal digits 8, 9, and this is a disadvantage which is unacceptable, as will appear below.

For k = 4, 2^k = 16, we can denote these tetrads of binary digits by the decimal digits 0, ..., 9 and six other symbols; e.g., the letters a, ..., f. It is best to let each decimal digit 0, .., 9 correspond to its binary expression (as an integer). Each letter a, ..., f may be treated as if it were the corresponding number in the sequence 10, ..., 15, and then it can be replaced by the binary expression of that number (as an integer). Thus we have to use the decimal digits 0, ..., 9 and the letters a, ..., f together, which may be a disadvantage: This disadvantage, however, is outweighed by the advantage that a printer which is organized in this fashion can print all decimal digits 0, ..., 9. It can also print the letters a, ..., f, which are likely to be useful for other indications.

We choose accordingly k = 4. The symbols that we assign to the 2^k = 16 tetrads of binary digits are shown in Table IV.

TABLE IV

0000	0	0100	4	1000	8	1100	c
0001	1	0101	5	1001	9	1101	d
0010	2	0110	6	1010	a	1110	e
0011	3	0111	7	1011	b	1111	f

In actual engineering terms this decision means the following things:

First: The typewriter which is used to produce information manually on the wire or tape has 16 keys, marked 0, ..., 9 and a, ..., f. If any one of these keys is depressed, it produces on the wire or tape the corresponding tetrad of binary digits, as indicated in Table IV.

Second: The printer which is used to convert information on the wire or tape into readable print, can print 16 symbols, the decimal digits 0, ..., 9 and the letters a, ..., f. It picks up the information on the wire, which is in the form of binary digits, tetradwise, and upon having sensed any tetrad, it prints the corresponding symbol, as indicated in Table IV.

Hence, a 40 binary digit word, e.g. a full size number, corresponds to 10 tetrads, i.e. to 10 symbols 0, ..., f; and a 20 binary digit half-word, e.g. an order, corresponds to 5 tetrads, i.e. to 5 symbols. Within an order the 12 first binary digits represent the location number x (cf. Table II), and therefore this x is separately represented by the 3 first tetrads, i.e. symbols.

It is best to allot to every decimally expressed number, too, the space of one word. This allows for 10 symbols, i.e. for 10 decimal digits. Since it is desirable, however, to leave space for a sign indication, these are not all available. The sign could be expressed by one binary digit, but it seems preferable not to disrupt the grouping in tetrads, and we propose therefore to devote an entire tetrad to the sign. In addition, it is desirable to denote the two signs + and - by a tetrad which differs in print from the decimal digits, i.e. which is printed in our system as a letter. We assign therefore tentatively the letter a (tetrad 1010) to + and the letter b (tetrad 1011) to -. Thus a 40 binary digit word can express a decimal number which consists of a sign and of 9 decimal digits. It is indicated to assume for the decimal numbers, as we did for the binary ones, that they lie between -1 and 1, i.e. that the decimal point follows immediately after the sign.

There is one more thing that need be mentioned in connection with the use of the sign indication in the decimal notation. Our binary notation handled negative numbers essentially by taking them modulo 2, i.e. a negative $a = -|a|$ was written as $2 + a = 2 - |a| = 1 + (1 - |a|)$. The first 1 expressed the - sign, and so the digits to the right of the binary point were those of $1 - |a|$, i.e. the complements of the digits of $|a|$ (with a correction of 1 in the extreme right place). (Cf. 5.7 in Part I of this report.) In our decimal notation it will prove more convenient to designate any number a (whether negative or not) by its sign indication followed (to the right of the decimal point) by the digits of $|a|$. This is in harmony with the usual notation in print, and it presents certain advantages in the actual process of conversion (cf. 9.6, 9.7).

9.5 The arithmetical consequences of these conventions are now easy to formulate.

The machine should be able to deal in two different ways with numbers that lie between -1 and 1 in terms of 40 binary digit words:

First: In the binary way: Sign digit and 39 binary digits.

Second: In the decimal or rather in the binary-decimal way: Sign tetrad, which is 1010 for + and 1011 for - (both aggregates to be read as binary numbers), and 9 decimal digits, each of which is expressed by a tetrad, that is simply its expression as a binary number.

In order to make this double system of notations effective, we must provide appropriate instructions for the conversions from each one to the other. I.e. our problem of conversion is to provide instructions for the binary to binary-decimal and the binary-decimal to binary conversions. It should be noted that these conversions take place between two 40 binary digit aggregates, and they are to be performed by our purely binary machine.

At this point a question concerning precision presents itself. We are considering conversions between what is effectively a 9 decimal digit number and between a 39 binary digit number, i.e. between a number given with the precision 10^{-9} and a number given with the precision $2^{-39} = 1.8 \cdot 10^{-12}$. The latter is 550 times more precise than the former, is this not a sign of unbalance? We wish to observe that this is not the case.

Indeed, the first mentioned, lower precision occurs in the decimal numbers, i.e. in the input (data) and the output (results); while the second mentioned, higher precision occurs in the binary numbers, i.e. in the intervening calculations. Hence the observed disparity in precisions means only that we carry in the calculations an excess precision factor of 550, i.e. of slightly less than 3 extra decimals, beyond the precision of the data and the results. Now it is clearly necessary to carry in any complicated calculation more decimals than the data or the results require, in order to keep the accumulation of the round-off errors down. For a calculation of such a complexity that justifies the use of a very high speed electronic computing device, 3 extra decimals will probably not be excessive from this point of view. They may even be inadequate, i.e. it may not be possible to utilize in the data and the results the full number (9) of available decimals.

These considerations conclude our preliminary discussion. We add only one more remark: With the exception of the arrangements relative to the typewriter and the printer (cf. Table IV and the two remarks which follow), we did not discuss physical features of the machine, but only arbitrary conventions, to be made effective by giving appropriate instructions to the machine. A variety of other conventions would be equally possible (e.g. different positioning of the decimal point, higher precision by using more than one word for a decimal and for a binary number, conversions into and from other systems than the decimal one, etc.), and could, if desired, be made effective by appropriate instructions. We are only discussing one example in this chapter, as a prototype, although it seems to be the simplest and the most immediately and generally useful one.

9.6 We are now in the position to discuss the actual conversion processes and their coding. Throughout this paragraph and the next one we will have to use alternately binary and decimal notations. We agree therefore that a digital sequence with no special markings is to be read decimally, e.g. 987016 or .75102; while one which is overscored is to be read binarily, e.g. $\overline{1011}$ or $.\overline{11001}$. Thus, e.g. $17 * \overline{10001}$, $.14 = .\overline{001000111101}\ldots$. In the tetrad notation of Table IV the two latter binary numbers would read 11 and .23d ... .

We consider first:

Problem 6.

The number a is stored, in its binary form, at a given memory location. It is desired to produce its binary-decimal form a', and to store it at another, given, memory location. ---

According to this a is a binary number, lying between -1 and 1. As pointed out at the end of 9.4, we must first sense the sign of a, and then convert |a|. |a| lies between 0 and 1, and its decimal expansion can be obtained by this inductive method:

(1) $a_0 = |a|$,

(2) $a_{i+1} + z_{i+1} = 10\, a_i,\ 0 \leq a_{i+1} < 1,\ z_{i+1} = 0, 1, \ldots, 9.$

If z^* is the sign indication of a, then the decimal expansion of a (cf. the remark at the end of 9.4) is $z^*, z_1, z_2, \ldots$. If we write $z_1, z_2, \ldots$ as (4 digit) binary numbers, then this representation becomes binary-decimal.

If $z^* \begin{cases} = \overline{1010} = 10 \text{ for } a \geq 0 \\ = \overline{1011} = 11 \text{ for } a < 0 \end{cases}$, then $z^*, z_1, z_2, \ldots, z_9$ is a sequence of 40 binary digits. Let us view it as a binary number between -1 and 1, in the sense of the ordinary conventions of our machine. Since z^* begins at any rate with 1, it will be negative, and z^* represents the actual number z^{**}

$$\begin{cases} = \overline{1.010} \text{ (binary, modulo 2)} = -1 + 2^{-2} = -2^{-2}\cdot 3 \text{ for } a \geq 0 \\ = \overline{1.011} \text{ (binary, modulo 2)} = -1 + 2^{-2} + 2^{-3} = -2^{-3}\cdot 5 \text{ for } a < 0 \end{cases}.$$

It is convenient to introduce

$$z_0 = 2^3\ z^{**} \begin{cases} = -6 \text{ for } a \geq 0 \\ = -5 \text{ for } a < 0 \end{cases}.$$ z_i represents the actual number $2^{-(4i+3)} z_i$. Hence the binary number that we want to produce is $a' = 2^{-3} z_0 + 2^{-7} z_1 + \ldots + 2^{-39} z_9$. This can be defined inductively as follows:

(3) $a'_0 = 2^{-39} z_0$, $z_0 \begin{cases} = -6 \text{ for } a \geq 0 \\ = -5 \text{ for } a < 0 \end{cases}$,

(4) $a'_{i+1} = 2^4 a'_i + 2^{-39} z_{i+1}$,

(5) $a' = a'_9$.

This completes our definitions, except that it is more convenient to write, from the point of view of our actual arithmetical processes, (2) in this form:

(2') $a_{i+1} + z_{i+1} = 2^4(2^{-1} a_i + 2^{-3} a_i),\ 0 \leq a_{i+1} < 1,\ z_{i+1} = 0, 1, \ldots, 9.$

This form of (2') means that the addition that is required in (2') is performed entirely within the range between -1 and 1, indeed between 0 and 1, and that the split into a "fractional part" a_{i+1} and an "integer part" z_{i+1}, i.e. the operation which necessarily requires exceeding that range, is initiated by a multiplication by 2^4 alone, i.e. by a quadruple left shift. This latter operation, a fourfold application of the L of Table II, automatically separates the fractional part and the integer part, referred to above: The former appears in the accumulator, the latter, multiplied by 2^{-39}, in the register. At the same time it should be noted, that (4) involves no true addition, but that if a_i' is in the register, the above mentioned fourfold application of L of Table II will replace it by $2^4 a_i'$ and will at the same time introduce there $2^{-39} z_{i+1}$, too, thereby forming the desired $a_{i+1}' = 2^4 a_i' + 2^{-39} z_{i+1}$. These considerations show, by the way, that it is best to form the successive values of a_i' (for i = 0, 1, ..., 9) in the register, and those of a_i in the accumulator. Since the register will not be required for any other purpose, we will even find it possible to store the successive values of a_i' in the register, while additional demands that will have to be made on the accumulator will make it impossible to do the same with a_i and the accumulator.

The iteration index i might be stored in the form $2^{-39} i$, but it is somewhat preferable to store it in the form $2^{-39}(8-i)$. (A corresponding trick would have been equally applicable in coding Problem 3.)

We can now draw the flow diagram, as shown in Figure 9.1. We note that the storage space indicated for a_i', B.1, is actually the register (cf. above), hence B.1 will not appear as such in the coding that follows. As to the final storage of the result, a', various alternatives are possible, including leaving it in the register. We put it instead into a definite memory location, and choose for this B.2.

FIGURE 9.1

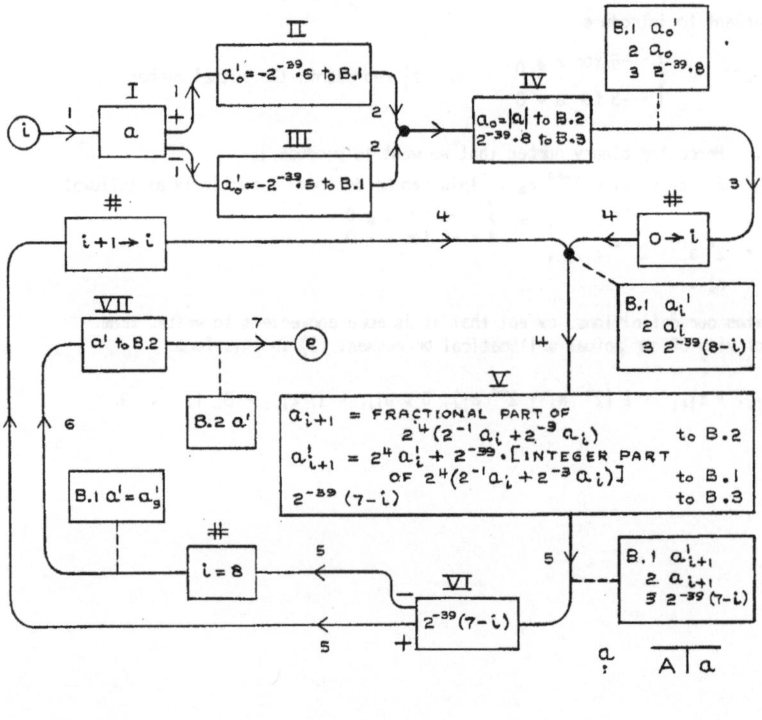

The boxes I-VII require static coding. We carry this out as in the preceding problems:

I,1	A		Ac	
2	II,1	Cc		
(to III,1)				
C.1	$-2^{-39}\cdot 6$			
II,1	C.1	R	R	$a_0' = -2^{-39}\cdot 6$
(to IV,1)				
C.2	$-2^{-39}\cdot 5$			
III,1	C.2	R	R	$a_0' = -2^{-39}\cdot 5$
(to IV,1)				
IV,1	A	M	Ac	a_0
2	B.2	S	B.2	a_0
C.3	$2^{-39}\cdot 8$			
IV,3	C.3		Ac	$2^{-39}\cdot 8$
4	B.3		B.3	$2^{-39}\cdot 8$
(to V,1)				
V,1	B.2		Ac	a_i
2		R	Ac	$2^{-1}a_i$
3	B.2	S	B.2	$2^{-1}a_i$
4		R		
5		R	Ac	$2^{-3}a_i$
6	B.2	h	Ac	$2^{-1}a_i + 2^{-3}a_i$
7		L		
8		L		
9		L		
10		L	Ac	a_{i+1} *)
			R	a'_{i+1} *)
11	B.2		B.2	a_{i+1}
12	B.3		Ac	$2^{-39}\cdot(8-i)$
C.4	2^{-39}			
V,13	C.4	h-	Ac	$2^{-39}\cdot(7-i)$
14	B.3	S	B.3	$2^{-39}\cdot(7-i)$
(to VI,1)				
VI,1	B.3		Ac	$2^{-39}\cdot(7-i)$
2	VII,1	Cc		
(to V,1)				
VII,1		A	Ac	$a' = a_9'$
2	B.2	S	B.2	a'
3	e	C		

Next, we order the boxes, as in the preceding problems. The ordering I, III, IV, V, VII; II; results, with two imperfections: IV, V should have been immediate successors of II, VI, respectively. This necessitates two extra orders

II,2	IV,1	C
VI,3	V,1	C

*) Cf. the definitions of these quantities in V, Figure 9.1.

We now must assign A, B.2–3, C.1–4 their actual values, pair the 29 orders I.1-2, II.1-2, III,1, IV,1-4, V,1-14, VI,1-3, VII,1-3 to 15 words, and then assign I,1 - VII,3 their actual values. These are expressed in this table:

I,1-2	0 - 0'	V,1-14	3'-10	II,1-2	13'-14
III,1	1	VI,1-3	10'-11'	A	15
IV,1-4	1'- 3	VII,1-3	12 -13	B.2-3	16 -17
				C.1-4'	18 -21

Now we obtain this coded sequence

0	15 , 13 Cc'	8	L , 16 S	15	a
1	19 R , 15 M	9	17 , 21 h-	16	-
2	16 S , 20	10	17 S , 17	17	-
3	17 S , 16	11	12 Cc, 3 C'	18	$-2^{-39}\cdot 6$
4	R , 16 S	12	A , 16 S	19	$-2^{-39}\cdot 5$
5	R , R	13	e C , 18 R	20	$2^{-39}\cdot 8$
6	16 h ,, L	14	1 C', - -	21	2^{-39}
7	L L				

The durations may be estimated as follows:

I: 75 μ, II: 75 μ, III: 50 μ, IV: 150 μ, V: 385 μ, VI: 125 μ, VII: 105 μ.
Total: I + (II or III) + IV + (V + VI) $\times$ 9 + VII =
= (75 + 75 + 150 + (385 + 125) $\times$ 9 + 105) μ = 4995 $\mu \approx$ 5 m .

9.7 We consider next:

Problem 7.

The number a' is stored, in its binary-decimal form, at a given memory location. It is desired to produce its binary form a, and to store it at another, given, memory location. —

According to this a' is a binary-decimal number, lying between -1 and 1. As pointed out at the end of 9.4, we must first sense the sign of a', which is determined by the digits 1-4 from the left, and then convert $|a'|$, which is directly given by the remaining digits 5-40 from the left. $|a'|$ lies between 0 and 1 and its binary expansion can be obtained as follows:

The digits 4i+1 to 4i+4 from the left in $|a'|$, i.e. in a' (cf. above), are the binary expression of the decimal digit w_i, which is digit i from the left in the decimal expansion of a' (i = 1, ..., 9). Hence we have the equation $|a'| = \Sigma_{i=1}^{9} 10^{-i} w_i$ which may be interpreted as a relation that is valid in the binary system. Hence it is better to write it like this: $|a| = \Sigma_{i=1}^{9} 10^{-i} w_i$.

Considering the precise definition of the operation L, the digits 4i+1 to 4i+4 (from the left) of a' can be made to appear as the digits 37-40 (from the left) in the register, if a' is placed into the accumulator and subjected to 4i+3 operations L. If we treat a' as if it were a binary number, this amounts to performing these operations:

(1) $a_0' + w^* = 2^3 a', \; 0 \leq a_0' < 1, \; w^* = 0, 1, \ldots, 7,$

(2) $a_{i+1}' + w_{i+1} = 2^4 a_i', \; 0 \leq a_{i+1}' < 1, \; w_{i+1} = 0, 1, \ldots, 15.$

w^* will correspond to the last three digits of the sign tetrad, i.e., $\overline{010} = 2$ or $\overline{011} = 3$, according to whether the number represented by a' (in its binary-decimal interpretation) is ≥ 0 or < 0. It is, however, simpler to get the sign indication from a' itself: If a' is interpreted as a binary number, then we have

$$a' \begin{cases} = \overline{1.010} \ldots {}^{*}) & < \\ = \overline{1.011} \ldots {}^{*}) & \geq \end{cases} \quad \overline{1.011} \; {}^{*}) = -1 + 2^{-2} + 2^{-3} = -2^{-3} \cdot 5$$

according to whether the number represented by a' (in its binary-decimal interpretation) is ≥ 0 or < 0. That is, this sign is the opposite of the sign of $a' + 2^{-3} \cdot 5$ in its binary interpretation.

$w_1, \ldots, w_9$ appear always as digits 37-40 (from the left) in the register; i.e., actually as the numbers $2^{-39} w_1, \ldots, 2^{-39} w_9$. It is therefore preferable to write the equation $|a| = \Sigma_{i=1}^{9} \; 10^{-i} \; w_i$ in this form: $|a| = \Sigma_{i=1}^{9} \; 10^{-i} \; 2^{39} \cdot 2^{-39} \; w_i$. In order to concentrate the divisions in one place we write $|a| = (\Sigma_{i=1}^{9} \; 10^{9-i} \cdot 2^{-39} \; w_i) : 2^{-39} \; 10^9$. This can be stated inductively as follows:

(3) $a_1 \qquad 2^{-39} \; w_1$

(4) $a_{i+1} = 10 \; a_i + 2^{-39} \; w_{i+1}$

(5) $a = a_9 : 2^{-39} \; 10^9.$

In fine a obtains from $|a|$ by noting that the sign of a is the opposite of the sign of $a' + 2^{-3} \cdot 5$ (binary interpretation, cf. above).

Again, $10 \; a_i$ is best formed as $2^3 \; a_i + 2 \; a_i$.

We can now draw the flow diagram, as shown in Figure 9.2. The induction index i can be treated as in the preceding problem.

The special use of the register which was made in Problem 6 is not possible here: a_i' has to be in the accumulator, since we want to form the fractional and integer parts of $2^4 \; a_i'$, and the operation L will only effect such things if we start from the accumulator; a_i has to go through the accumulator, since we have to form true sums (with carries) like $2^3 \; a_i + 2 \; a_i$; and neither can occupy the accumulator continuously, since the accumulator is needed alternately for both as well as for the sensing in connection with conditional transfer orders. There are, however, certain storage functions, which are best handled by the accumulator or register, in a manner not made explicit in the flow diagram. Thus $2^{-39} \cdot (7-i)$ is best stored in the accumulator as well as in B.3 before the use in III; $|a|$ is best stored in the register after IV, and the move of a to B.2 is only made subsequently in VI and in VII.

---

*) Binary, reduced modulo 2 into the interval -1, 1.

FIGURE 9.2

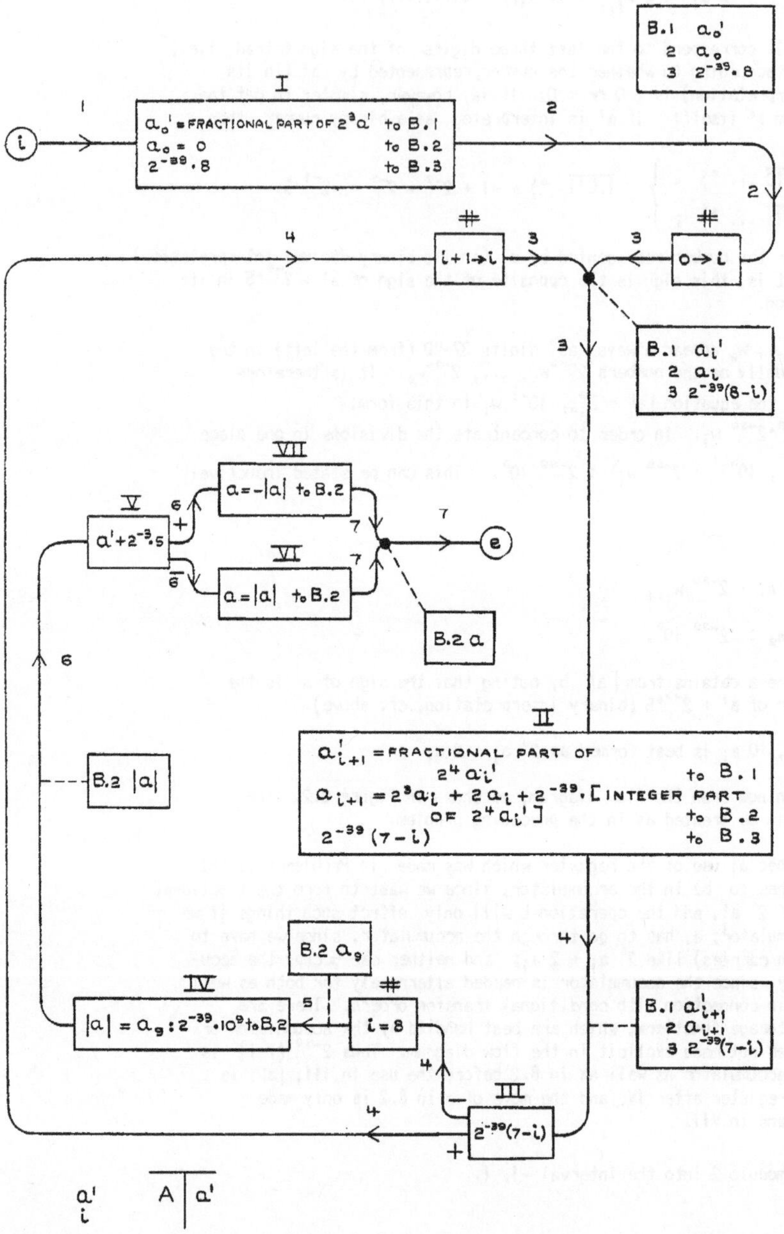
I
a_0' = FRACTIONAL PART OF $2^3a'$ to B.1
$a_0 = 0$ to B.2
$2^{-39} \cdot 8$ to B.3
B.1 a_0'
2 a_0
3 $2^{-39} \cdot 8$
#
$i+1 \to i$
#
$0 \to i$
B.1 a_i'
2 a_i
3 $2^{-39}(8-i)$
VII
$a = -|a|$ to B.2
V
$a' + 2^{-3} \cdot 5$
VI
$a = |a|$ to B.2
B.2 a
B.2 |a|
II
a_{i+1}' = FRACTIONAL PART OF $2^4 a_i'$ to B.1
$a_{i+1} = 2^3 a_i + 2a_i + 2^{-39} \cdot$ [INTEGER PART OF $2^4 a_i'$] to B.2
$2^{-39}(7-i)$ to B.3
B.2 a_9
IV
$|a| = a_9 : 2^{-39} \cdot 10^9$ to B.2
#
$i = 8$
B.1 a_{i+1}'
2 a_{i+1}
3 $2^{-39}(7-i)$
III
$2^{-39}(7-i)$
a_i'
A | a'

The boxes I-VII require static coding. We carry this out as in the preceding problems:

I.1			Ac	a'
2		L		
3		L		
4		L	Ac	a'_0 *)
5	B.1	S	B.1	a'_0
C.1	0			
I,6	C.1		Ac	0
7	B.2		B.2	0
C.2	$2^{-39}\cdot 8$			
I.8	C.2		Ac	$2^{-39}\cdot 8$
9	B.3		B.3	$2^{-39}\cdot 8$
(to II,1)				
II,1	B.1		Ac	a'_i
2		L		
3		L		
4		L		
5		L	Ac	a'_{i+1} *)
			R	$2^{-39}w_{i+1}$ *)
6	B.1		B.1	a'_{i+1}
7			Ac	$2^{-39}w_{i+1}$
8	s.1		s.1	$2^{-39}w_{i+1}$
9	B.2		Ac	a_i
10		L		
11		L	Ac	$2^2 a_i$
12	B.2	h	Ac	$2^2 a_i + a_i$
13		L	Ac	$2^3 a_i + 2a_i$
14	s.1	h	Ac	a_{i+1} *)
15	B.2	S	B.2	a_{i+1}
16	B.3		Ac	$2^{-39}\cdot(8-i)$
C.3	2^{-39}			
II,17	C.3	h-	Ac	$2^{-39}\cdot(7-i)$
18	B.3	S	B.3	$2^{-39}\cdot(7-i)$
(to III,1)				
III,1	II,1	Cc		
(to IV,1)				
IV,1	B.2		Ac	a_9
C.4	$2^{-39}\ 10^9$			
IV,2	C.4		R	$\lvert a\rvert = a_9 : 2^{-39}10^9$
(to V,1)				
V,1	A		Ac	a'
C.5	$2^{-3}\cdot 5$			
V,2	C.5	h	Ac	$a' + 2^{-3}\cdot 5$
3	VII,1	Cc		
(to VI,1)				

*) Cf. the definitions of these quantities in I, II, Figure 9.2, and in (1) - (5).

VI,1 (to VII,4)		.	Ac	a = |a|
VII,1		A	Ac	|a|
2	B.2	S	B.2	|a|
3	B.2	-	Ac	a = - |a|
4	B.2	S	B.2	a
5	e	C		

Next, we order the boxes, as in the preceding problems. The ordering I, II, III, IV, V, VI; VII results, with one imperfection, VII,4 should have been the immediate successor of VI. This necessitates one extra order

VI,2	VII,4	C	

We must now assign A, B.1-3, C.1-5, s.1 their actual values, pair the 40 orders I,1-9, II,1-18, III,1, IV,1,2, V,1-3, VI,1-2, VII,1-5 to 20 words, and then assign I,1 - VII,5 their actual values. We wish to do this as a continuation of the code of 9.6. We will therefore begin with the number 22. Furthermore the contents of C.2,3 coincide with those of 20, 21 in the code of 9.6, and the contents of B.1-3 and of s.1 are irrelevant like those of 16, 17 there. Hence two of these may be made to coincide with 16, 17. Also a is in 15 and a' is produced in 16 there, while we have now a' in A and produce a in B.2. Hence, we identify our A, B.1,2 with 16, 17, 15 there. Summing all these things up, we obtain the following table:

I,1-9	22 - 26	VI,1-2	38' - 39	B.3	42
II,1-18	26' - 35	VII,1-5	39'-41'	C.1	43
III,1	35'	A	16	C.2-3	20-21
IV,1-2	36 - 36'	B.1	17	C.4-5	44-45
V,1-3	37 - 38	B.2	15	s.1	46

Now we obtain this coded sequence:

22	16 , L	30	46 S , 15	38	41 Cc, A
23	L , L	31	L , L	39	41 C , A
24	17 S , 48	32	15 h , L	40	15 S , 15 -
25	15 S , 20	33	46 h , 15 S	41	15 S , e C
26	42 S , 17	34	42 , 21 h-	42	- - - -
27	L , L	35	42 S , 36 Cc	43	0
28	L , L	36	15 , 44 ÷	44	2^{-39} 10^9
29	17 S , A	37	16 , 45 h	45	$2^{-3} \cdot 5$
				46	- - - -

The durations may be estimated as follows:

I: 290 μ, II: 515 μ, III: 50 μ, IV: 195 μ, V: 125 μ, VI: 55 μ, VII: 180 μ.

Total: I + (II + III) × 9 + IV + V + ((VI + VII 4,5) or VII) =
= (290 + (515 + 50) × 9 + 195 + 125 + 180) μ = 5,875 μ ≈ 5,9 m.

We may summarize the results of 9.6 and 9.7 as follows:

The binary-decimal and the decimal-binary conversions can be provided for by instructions which require 47 words, i.e. 1.1% of the total (selectron) memory capacity of 4,096 words. These conversions last 5 m and 5.9 m, respectively. This is short compared to the 100 m which, as was shown in 9.3, is a reasonable standard for conversion durations.

9.8 We proceed now to the problems of the second class referred to in 9.1 and 9.2: The problems that arise when we undertake to operate the machine in such a manner, that it deals effectively with numbers of more than 40 binary digits.

There are various ways to deal with this situation, and it is actually advisable to study every problem which requires more than normal (40 binary digit) precision as an individual case: The most efficient way to keep track of numerical effects beyond the 40 binary digit limit may vary considerably from case to case. On the other hand, our interest is at present more in giving examples of our methods of coding in typical situations, than in devising particularly efficient ways to handle cases that require more than normal precision. We will therefore restrict our discussion to the most straightforward procedure, although more flexibility and speed could probably be secured in many actual instances by different schemes.

The procedure to which we alluded is that described in 9.2: By using k words to express a number, we obtain numbers with (a sign and) 39 k binary digits. We will actually restrict ourselves to the case of k = 2: Two words per number of (a sign and) 78 binary digits. In this way the precision becomes $2^{-78} = 3.3 \cdot 10^{-24}$, or (considering that the length of the available interval -1, 1 is 2) $2^{-79} = 1.6 \cdot 10^{-24}$

We will store every number in two successive words: The first word contains the sign of the number in question and its digits 1-39 (from the left); in the second word we put the sign digit equal to zero, and then let the digits 40-78 (from the left) of the number follow.

The operations which have to be discussed are the four basic operations of arithmetics: Addition, subtraction, multiplication, division.

Let us consider addition and subtraction first:

Problem 8.

The (double precision) numbers a, b are stored in two given pairs of memory locations. Form $a \pm b$ (also double precision) and store it at another given, pair of memory locations. —

Let a be stored at A.1,2, b at A.3,4, and direct $a \pm b$, when formed, back to A.1,2. Denote the first and the second parts of a, b, $a \pm b$ by $\bar{a}$, $\bar{b}$, $\overline{a \pm b}$ and $\bar{\bar{a}}$, $\bar{\bar{b}}$, $\overline{\overline{a \pm b}}$, respectively.

It is easily seen that

$$(1) \qquad \overline{\overline{a \pm b}} = \bar{\bar{a}} \pm \bar{\bar{b}} + \varepsilon ,$$

where ε is the sign digit of $\bar{\bar{a}} \pm \bar{\bar{b}}$,

$$(2) \qquad \overline{a \pm b} = \bar{a} \pm \bar{b} \pm 2^{-39}\varepsilon .$$

It should be remembered, of course, that all these relations must be interpreted modulo 2, and that the positional value of the sign digit is 2^0. Since $\bar{\bar{a}}, \bar{\bar{b}}$ have the sign digit 0, a sign digit 1 in $\bar{\bar{a}} \pm \bar{\bar{b}}$ is equivalent to a carry from digit 40 to digit 39 (from the left).

We can now draw the flow diagram, as shown in Figure 9.3. It is convenient to carry out some storage functions in a manner not made explicit in the flow diagram. Thus $\overline{\overline{a \pm b}}$ is best formed and moved into A.2 already in the course of II and III, and while $\overline{a \pm b}$ is built up in the accumulator in the course of IV, the ground for this can be effectively prepared (in the accumulator) towards the end of II and III. These arrangements have the effect, that neither B.1 nor B.2 is actually required, both represent storage effected at appropriate times in the accumulator.

FIGURE 9.3

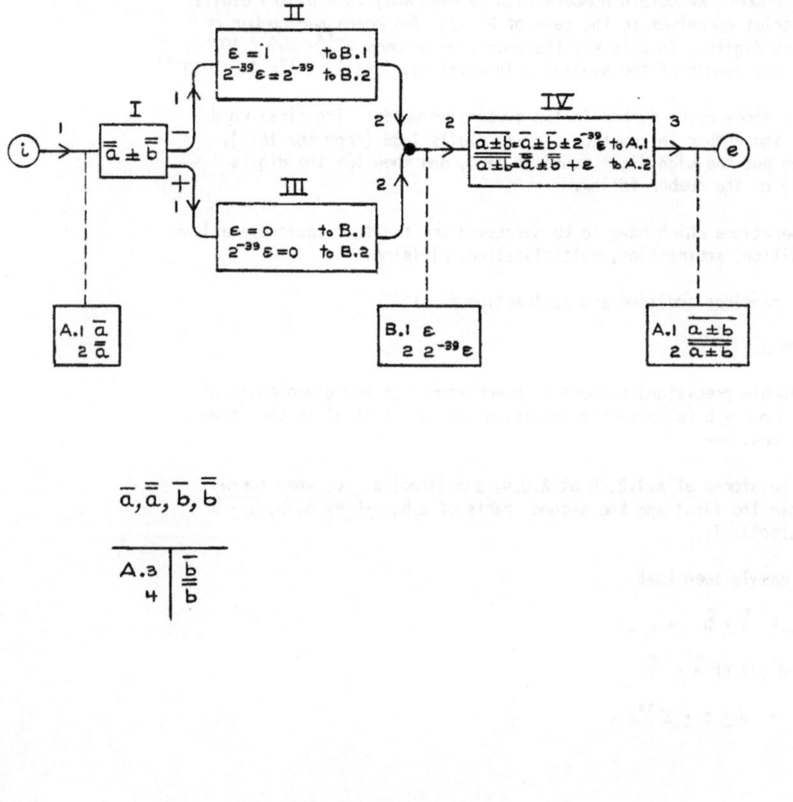

The static coding of the boxes I-IV follows. Those orders which depend on the alternative $\pm$, are marked ϕ in their + form and $\phi\phi$ in their − form.

I-1	A.2			Ac	$\overline{\overline{a}}$
2 ϕ	A.4	h		Ac	$\overline{\overline{a \pm b}}$
2 $\phi\phi$	A.4	h−		Ac	$\overline{\overline{a - b}}$
3	III-1	Cc			
(to II,1)					
C.1	−1				
II,1	C.1	h−		Ac	$\overline{\overline{a \pm b}}$ *)
2	A.2	S		A.2	$\overline{a \pm b}$
C.2	2^{-39}				
II,3 ϕ	C.2			Ac	$2^{-39}\varepsilon = 2^{-39}$
3 $\phi\phi$	C.2			Ac	$-2^{-39}\varepsilon = -2^{-39}$
(to IV,1)					
III,1	A.2			A.2	$\overline{a \pm b}$ *)
C,3	0				
III,2	C,3			Ac	0
(to IV,)					
IV,1	A.1	h		Ac	$\overline{a} \pm 2^{-39}\varepsilon$
2 ϕ	A.3	h		Ac	$\overline{a + b}$ *)
2 $\phi\phi$	A.3	h−		Ac	$\overline{a - b}$ *)
3	A.1	S		A.1	$\overline{a \pm b}$
4	e	C			

The ordering of the boxes is I, II, IV; III, and IV must also be the immediate successor of III. This necessitates the extra order

III,3   IV.1

We must now assign A.1-4, C.1-3, their actual values, pair the 13 orders I,1-3, II,1-3, III,1-3, IV,1-4 to 7 words, and then assign I,1 − IV.4 their actual values. These are expressed in this table:

I,1-3	0 − 1	IV,1-4	3 − 4'	A.1-4	7 − 10
II, 1-3	1'-2'	III,1-3	5 − 6	C.1-3	11 − 13

Now we obtain this coded sequence:

0 ϕ	8	,	10 h	3 ϕ	7 h,	9 h	8	$\overline{\overline{a}}$
0 $\phi\phi$	8	,	10 h−	3 $\phi\phi$	7 h,	9 h−	9	$\overline{\overline{b}}$
1	5 Cc	,	11 h−	4	7 S,	e C	10	b
2 ϕ	8 S	,	12	5	8 S,	13	11	−1
2 $\phi\phi$	8 S		12 −	6	3 C,	− −	12	2^{-39}
				7	$\overline{\overline{a}}$		13	0

*) Cf. the definitions of these quantities in IV, Figure 9.3.

If both variants (+ and -) are to be stored in the memory (which is likely to be the case, if they are needed at all), their 7-13 can be common. Indeed 11-13 are likely to occur in other coded sequences, too. So the two variants are likely to require together 18 to 21 words.

The durations may be estimated as follows:

I: 125 μ, II: 125 μ, III: 125 μ, IV: 150 μ.
Total: I + (II or III) + IV = (125 + 125 + 150) μ = 400 μ = .4 m.

To sum up:

The double precision addition and subtraction can be provided for by instructions which require 18 to 21 words, i.e. about 5% of the total (selectron) memory capacity. The operations last 400 μ = .4 m each.

These sequences are comparable to A.I, A.3 h (or A.3 h-), A.I S in the case of ordinary precision, which consume 125 μ. Thus the double precision slows addition and subtraction by a factor 3.2. The factor is actually slightly higher, because the double precision system, as described above, is less flexible than the ordinary one. However, as mentioned previously, there are various ad hoc methods to increase its flexibility, but we do not propose to discuss them here.

9.9 We consider now the multiplication:

Problem 9.

The (double precision) numbers a, b are stored in two given pairs of memory locations. Form ab (also double precision) and store it at another, also given, memory location.---

As in 9.8, let a be stored at A.I,2, b at A.3,4. It is more convenient to direct ab, when formed, to A.3,4. Denote again the first and the second parts of a, b, ab by $\bar{a}$, $\bar{b}$, $\overline{ab}$ and $\bar{\bar{a}}$, $\bar{b}$, $\overline{\overline{ab}}$, respectively.

The multiplication order (II, Table II) multiplies two 39 digit quantities u, v by forming a 78 digit product uv. The sign and the digits I-39 are formed in the accumulator, while the digits 40-78 (with a sign digit 0) are formed in the register. These are, of course, the $\overline{uv}$ and $\overline{\overline{uv}}$ respectively, as we defined them above. They are, however, subject to a subsequent (round off) modification, as described in, and discussed in connection with, the multiplication order referred to above. We denote therefore the actual contents of the accumulator and of the register by $\overrightarrow{\overline{uv}}$ and by $\overrightarrow{\overline{\overline{uv}}}$. $\overrightarrow{\overline{uv}}$ is, of course, the quantity which we denoted by uv in all past discussions (as well as in all future discussions, outside this paragraph), the product rounded to 39 digits, stored in the accumulator. Now, however, it is necessary to be more precise, and to discriminate between uv, $\overrightarrow{\overline{uv}}$, $\overrightarrow{\overline{\overline{uv}}}$.

It is clear, that $\overline{ab}$ is essentially $\overrightarrow{\bar{a}\bar{b}}$, and that $\overline{\overline{ab}}$ is essentially $\overrightarrow{\bar{a}\,\bar{\bar{b}}} + \overrightarrow{\bar{\bar{a}}\,\bar{b}} + \overrightarrow{\overrightarrow{\bar{a}\,\bar{b}}}$. The only things that these expressions do not account for, are the carries in ab from digit 40 to digit 39. It is easily seen, that these, too, are completely accounted for by the following formulae:

(1) $u = \overrightarrow{\bar{a}\,\bar{b}} + \varepsilon_1$,
where ε_1 is the sign digit of $\overrightarrow{\bar{a}\,\bar{b}}$,

(2) $v = \overrightarrow{\bar{\bar{a}}\,\bar{b}} + \varepsilon_2$,
where ε_2 is the sign digit of $\overrightarrow{\bar{\bar{a}}\,\bar{b}}$,

(3) $w = u + v + \varepsilon_3$,
where ε_3 is the sign digit of $u + v$,

(4) $z = \overrightarrow{\bar{a}\,\bar{\bar{b}}} + \varepsilon_4$,
where ε_4 is the sign digit of $\overrightarrow{\bar{a}\,\bar{\bar{b}}}$,

(5) $\overline{a\,b} = w + z + \varepsilon_5$,
where ε_5 is the sign digit of $w + z$,

(6) $\overline{a\,b} = \overrightarrow{\bar{a}\,\bar{b}} + 2^{-39}\,(-\varepsilon_1 - \varepsilon_2 + \varepsilon_3 - \varepsilon_4 + \varepsilon_5)$.

It should again be remembered, that all these relations must be interpreted modulo 2, and that the positional value of a sign digit is 2^0.

We can now draw the flow diagram, as shown in Figure 9.4. There will again be some transient storage in the accumulator, as well as in A.3,4. Since we have discussed similar situations in previous Problems, we need not go now into this more fully.

The static coding of the boxes I-XVI follows. There will be one deviation from the scheme indicated in Figure 9.4: $\overrightarrow{\bar{\bar{a}}\,\bar{b}}$ is formed in X, hence at the same time $\overrightarrow{\bar{a}\,\bar{b}}$ becomes available. Since $\overrightarrow{\bar{a}\,\bar{b}}$ will be needed in XVI, we will store it after X. Since $\bar{b}$ occupying A.3 is needed after X solely for the purpose of forming $\overrightarrow{\bar{\bar{a}}\,\bar{b}}$, we can remove $\bar{b}$, and store $\overrightarrow{\bar{a}\,\bar{b}}$ in A.3.

I,1	A.1	R	R	$\bar{a}$
2	A.4	x	Ac	$\overrightarrow{\bar{a}\,\bar{b}}$
3	II,1	Cc		
(to III,1)				
II,1	A.4		A.4	$\overrightarrow{\bar{a}\,\bar{b}}$
C.1	0			
II,2	C.1		Ac	0
3	B		B	0
(to IV,1)				
C.2	-1			
III,1	C.2	h-	Ac	$u = \overrightarrow{\bar{a}\,\bar{b}} + 1$
2	A.4	S	A.4	u
C.3	2^{-39}			
III,3	C.3		Ac	-2^{-39}
4	B		B	-2^{-39}
(to IV,1)				
IV,1	A.2	R	R	$\bar{\bar{a}}$
2	A.3	x	Ac	$\overrightarrow{\bar{\bar{a}}\,\bar{b}}$
3	V,1	Cc		
(to VI,1)				
V,1	A.4	h	Ac	$u + v = u + \overrightarrow{\bar{\bar{a}}\,\bar{b}}$
2	A.4	S	A.4	$u + v$
(to VII,1)				

FIGURE 9.4

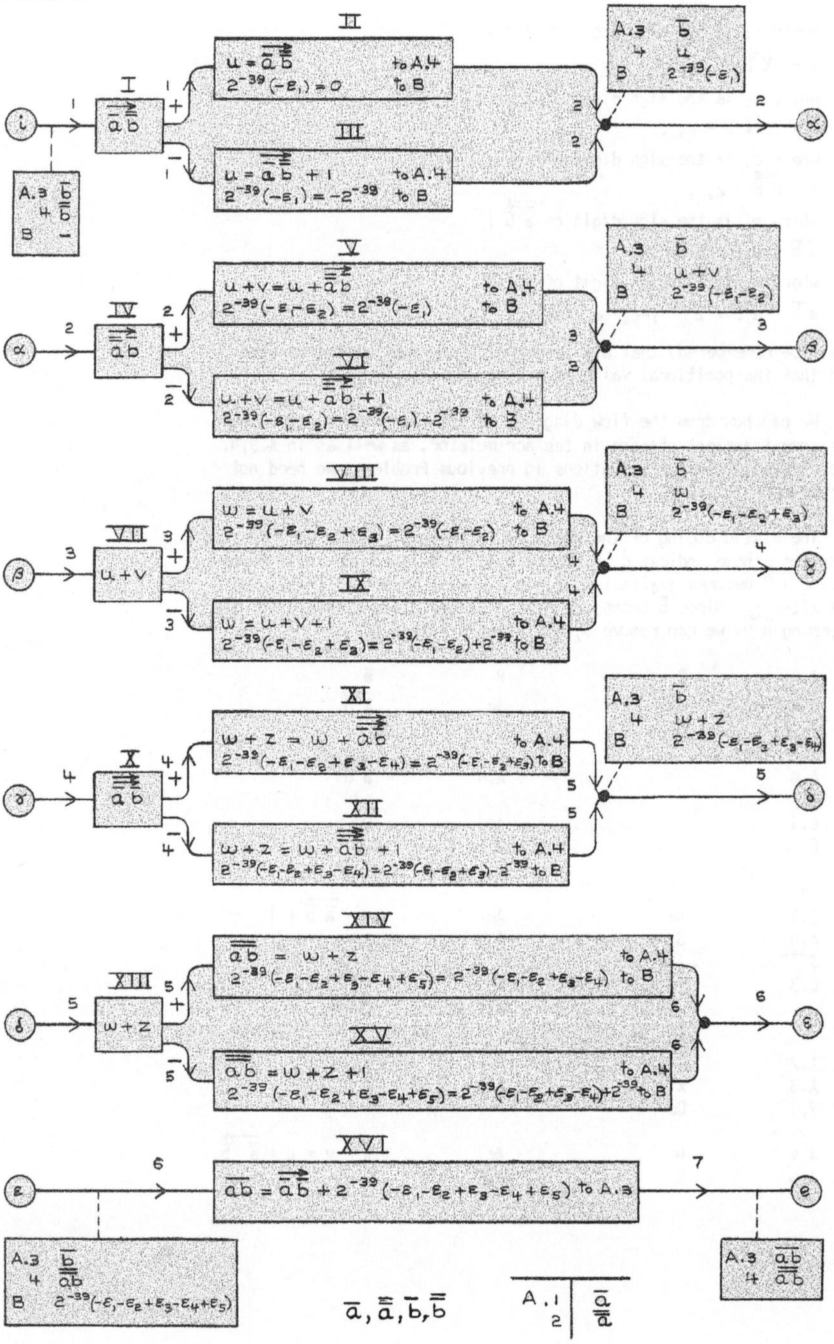
I
ab
A.3 b
4 b
B 1
II
u = ab to A.4
2^-39(-ε1) = 0 to B
III
u = ab + 1 to A.4
2^-39(-ε1) = -2^-39 to B
A.3 b
4 u
B 2^-39(-ε1)
IV
ab
V
u+v = u + ab to A.4
2^-39(-ε1-ε2) = 2^-39(-ε1) to B
VI
u+v = u + ab + 1 to A.4
2^-39(-ε1-ε2) = 2^-39(-ε1) - 2^-39 to B
A.3 b
4 u+v
B 2^-39(-ε1-ε2)
VII
u+v
VIII
w = u+v to A.4
2^-39(-ε1-ε2+ε3) = 2^-39(-ε1-ε2) to B
IX
w = u+v+1 to A.4
2^-39(-ε1-ε2+ε3) = 2^-39(-ε1-ε2) + 2^-39 to B
A.3 b
4 w
B 2^-39(-ε1-ε2+ε3)
X
ab
XI
w + z = w + ab to A.4
2^-39(-ε1-ε2+ε3-ε4) = 2^-39(-ε1-ε2+ε3) to B
XII
w + z = w + ab + 1 to A.4
2^-39(-ε1-ε2+ε3-ε4) = 2^-39(-ε1-ε2+ε3) - 2^-39 to B
A.3 b
4 w+z
B 2^-39(-ε1-ε2+ε3-ε4)
XIII
w+z
XIV
ab = w + z to A.4
2^-39(-ε1-ε2+ε3-ε4+ε5) = 2^-39(-ε1-ε2+ε3-ε4) to B
XV
ab = w + z + 1 to A.4
2^-39(-ε1-ε2+ε3-ε4+ε5) = 2^-39(-ε1-ε2+ε3-ε4) + 2^-39 to B
XVI
ab = ab + 2^-39(-ε1-ε2+ε3-ε4+ε5) to A.3
A.3 b
4 ab
B 2^-39(-ε1-ε2+ε3-ε4+ε5)
A.3 ab
4 ab
a, a, b, b
A.1 a
2 a

VI,1	A.4	h	Ac	$u + \overrightarrow{\bar{a}\,b}$
2	C.2	h-	Ac	$u + v = u + \overrightarrow{\bar{\bar{a}}\,b} + 1$
3	A.4	S	A.4	$u + v$
4	C.3	-	Ac	-2^{-39}
5	B	h	Ac	$2^{-39}(-\varepsilon_1)\ -2^{-39}$
6	B	S	B	$2^{-39}(-\varepsilon_1)\ -2^{-39}$
(to VII,1)				
VII,1	A.4		Ac	$u + v$
2	VIII,1	Cc		
(to IX,1)				
VIII	------			
(to X,1)				
IX,1	C.2	h-	Ac	$w = u + v$
2	A.4	S	A.4	w
3	C.3		Ac	2^{-39}
4	B	h	Ac	$2^{-39}(-\varepsilon_1 - \varepsilon_2) + 2^{-39}$
5	B	S	B	$2^{-39}(-\varepsilon_1 - \varepsilon_2) + 2^{-39}$
(to X,1)				
X,1	A.1	R	R	$\bar{a}$
2	A.3	×	Ac	$\overrightarrow{\bar{a}\,b}$
			R	$\overrightarrow{\overline{\bar{a}\,b}}$
3	A.3	-	A.3	$\overrightarrow{\bar{a}\,b}$
4		A	Ac	$\overrightarrow{\overline{\bar{a}\,b}}$
5	XI,1	Cc		
(to XII,1)				
XI,1	A.4	h	Ac	$w + z = w + \overrightarrow{\overline{\bar{a}\,b}}$
2	A.4	S	A.4	$w + z$
(to XIII,1)				
XII,1	A.4'	h	Ac	$w + \overrightarrow{\overline{\bar{a}\,b}}$
2	C.2	h-	Ac	$w + z = w + \overrightarrow{\overline{\bar{a}\,b}} + 1$
3	A.4	S	A.4	$w + z$
4	C.3	-	Ac	-2^{-39}
5	B	h	Ac	$2^{-39}(-\varepsilon_1 - \varepsilon_2 + \varepsilon_3)\ -2^{-39}$
6	B	S	B	$2^{-39}(-\varepsilon_1 - \varepsilon_2 + \varepsilon_3)\ +2^{-39}$
(to XIII,1)				
XIII,1	A.4		Ac	$w + z$
2	XIV,1	Cc		
(to XV,1)				
XIV	------			
(to XVI,1)				
XV,1	C.2	h-	Ac	$\overline{\overline{a\,b}} = w + z + 1$
2	A.4	S	A.4	$\overline{\overline{a\,b}}$
3	C.3		Ac	2^{-39}
4	B	h	Ac	$2^{-39}(-\varepsilon_1 - \varepsilon_2 + \varepsilon_3 - \varepsilon_4) + 2^{-39}$
5	B	S	Ac	$2^{-39}(-\varepsilon_1 - \varepsilon_2 + \varepsilon_3 - \varepsilon_4) + 2^{-39}$
(to XVI,1)				

XVI,1	A.3		Ac	$\overrightarrow{\bar{a}\,b}$
2	B	h	Ac	$\overline{a\,b} = \overrightarrow{\bar{a}\,b} + 2^{-39}(-\varepsilon_1 - \varepsilon_2 + \varepsilon_3 - \varepsilon_4 + \varepsilon_5)$
3	A.3	S	A.3	$\overline{a\,b}$
4	e	C		

Note, that the boxes VIII and XIV required no coding, hence their immediate successors (X,1 and XVI,1) must follow directly upon their immediate predecessors. However, these two boxes have actually no immediate predecessors, VIII,1 and XIV,1 appear in the Cc orders VII,2 and XIII,2. Hence they must be replaced there by X,1 and XVI,1.

The ordering of the boxes is I, III, IV, VI, VII, IX, X, XII, XIII, XV XVI; II; V; XI (VIII and XIV omitted, cf. above), and IV, VII, XIII must also be the immediate successors of II, V, XI, respectively. This necessitates the extra orders

II,4	IV,1	Cc
V,3	VII,1	Cc
XI,3	XIII,1	Cc

We must now assign A.1-4, B, C.1-3 their actual values, pair the 55 orders I,1-3, II,1-4, III,1-4, IV,1-3, V,1-3, VI,1-6, VII,1-2, IX,1-5, X,1-5, XI,1-3, XII,1-6, XIII,1-2, XV,1-5, XVI,1-4 to 28 words, and then assign I,1 - XVI,4 their actual values. These are expressed in this table:

I,1-3	0 - 1	IX,1-5	9 - 11	II,1-4	22'-24
III,1-4	1'- 3	X,1-5	11'-13'	V,1-3	24'-25'
IV,1-3	3'-4'	XII,1-6	14 -16'	XI,1-3	26 - 27
VI,1-6	5 - 7'	XIII,1-2	17 -17'	A.1-4	28 - 31
VII,1-2	8 - 8'	XV,1-5	18 - 20	B	32
		XVI,1-4	20'-22	C.1-3	33 - 35

Now we obtain this coded sequence:

0	28 R ,	31 x	12	30 x ,	30 S	24	3 Cc',	31 h
1	22 Cc',	34 h-	13	A ,	26 Cc	25	31 S ,	8 Cc
2	31 S ,	35 -	14	31 h ,	34 h-	26	31 h ,	31 S
3	32 S ,	29 R	15	31 S ,	35 -	27	17 Cc ,	- -
4	30 x ,	24 Cc'	16	32 h ,	32 S	28	$\bar{a}$	
5	31 h ,	34 h-	17	31 ,	20 Cc'	29	$\bar{\bar{a}}$	
6	31 S ,	35 -	18	34 h-,	31 S	30	$\bar{b}$	
7	32 h ,	32 S	19	35 ,	32 h	31	$\bar{\bar{b}}$	
8	31 ,	11 Cc'	20	32 S ,	30	32	-	
9	34 h-,	31 S	21	32 h ,	30 S	33	0	
10	35 ,	32 h	22	e C ,	31 S	34	-1	
11	32 S ,	28 R	23	33 ,	32 S	35	2^{-39}	

It is likely that this will be stored in the memory together with the two variants (+ and -) of Problem 8. Hence 28-31, 33-35 of this code can be common with 7-10, 13, 11, 12, respectively, in the code of that problem. Also, 32 is likely to occur in other coded sequences, too. So this problem is likely to require 28 or 29 additional words.

The durations may be estimated as follows:

I: 195 μ, II: 150 μ, III: 150 μ, IV: 195 μ, V: 125 μ, VI: 225 μ, VII: 75 μ, IX: 200 μ, X: 250 μ, XI: 125 μ, XII: 225 μ, XIII: 75 μ, XV: 200 μ, XVI: 150 μ.

Total: I + (II or III) + IV + (V or VI)+ VII + (nothing or IX) + X + + (XI or XII) + XIII + (nothing or XV) + XVI

average = (195 + 150 + 195 + 175 + 75 + 100 + 250 + 175 + 75 + + 100 + 150) μ = 1,640 μ = 1.6 m.

To sum up:

The double precision multiplication can be provided for by instructions which require 28 or 29 words in addition to those required by Problem 8 (double precision addition and subtraction) i.e. about an additional .7% of the total (selectron) memory capacity. The operation lasts 1,640 μ ≈ 1.6 m.

This sequence is comparable to A.1 R, A.3 x, A.3 S in the case of ordinary precision, which consume 195 μ. Thus the double precision slows multiplication by a factor 8.4. In contrast with the situation described in connection with the double precision addition and subtraction (cf. the end of 9.8), there is probably no significant loss of flexibility involved in the present case.

9.10 Division offers no particular difficulties. In order to obtain a double precision quotient of a (represented by $\bar{a}$, $\bar{\bar{a}}$) by b (represented by $\bar{b}$, $\bar{\bar{b}}$) one may first "approximate" the desired double precision quotient $q = \frac{a}{b}$, by forming the ordinary precision quotient $q_o = \frac{\bar{a}}{\bar{b}}$, and then refine the result by any one of several well known methods. The availability of double precision addition, subtraction, and multiplication makes these matters quite simple. We do not propose, however, to go now further into this subject.

We may summarize the results of 9.8 and 9.9 as follows:

Double precision arithmetics, comprising addition, subtraction and multiplication, can be provided for by instructions which require 46 to 50 words, i.e. 1.2% of the total (selectron) memory capacity. Due to these changes addition and subtraction are slowed by a factor 3.2 and multiplication by a factor 8.4, in comparison to ordinary precision operations. Furthermore there is a certain loss in the flexibility of the use of the two first mentioned operations. It seems therefore reasonable to expect an overall, average slowing by a factor between 6 and 8 in comparison to ordinary precision operations.

PLANNING AND CODING OF PROBLEMS
FOR AN
ELECTRONIC COMPUTING INSTRUMENT

By

Herman H. Goldstine John von Neumann

Report on the Mathematical and Logical Aspects of an
Electronic Computing
Instrument

Part II, Volume II

The Institute for Advanced Study
Princeton, New Jersey
1948

PREFACE

This report was prepared in accordance with the terms of Contract No. W-36-034-ORD-7481 between the Research and Development Service, U. S. Army Ordnance Department and The Institute for Advanced Study. It is a continuation of our earlier report entitled, "Planning and Coding of Problems for an Electronic Computing Instrument", and it constitutes Volume II of Part II of the sequence of our reports on the electronic computer. Volume III of Part II will follow within a few months.

Herman H. Goldstine

John von Neumann

The Institute for Advanced Study
15 April 1948

TABLE OF CONTENTS

INTRODUCTORY REMARK

We found it convenient to make some minor changes in the positioning of the parts of all orders of Table II, and in the specific effect of two among them.

First: We determine that in each order of Table II the memory position number x occupies not the 12 left-hand digits (cf. the second remark of 7.2, cf. also 8.2), but the 12 right-hand digits. I.e. the digits 9 to 20 or 29 to 40 (from the left) for a left-hand or a right-hand order, respectively.

Second: We change the orders 18, 19 (the partial substitution orders xSp, xSp'), so that they substitute these new positions, and these crosswise. I.e. 18 replaces the 12 right-hand digits of the left-hand order located at x (i.e. the digits 9 to 20 [from the left]) by the 12 digits 29 to 40 (from the left) in the accumulator. Similarly 19 replaces the 12 right-hand digits of the right-hand order located at x (i.e. the digits 29 to 40 [from the left]) by the 12 digits 9 to 20 (from the left) in the accumulator.

For the standard form of a position mark, $x_0 = 2^{-19}x + 2^{-39}x$, as introduced in 8.2, these changes compensate each other and have no effect. Therefore all the uses that we made so far of these orders are unaffected.

On the other hand the new arrangement permits certain arithmetical uses of these orders, i.e. uses when the position x is not occupied by orders at all, but when it is used as (transient) storage for numbers. In this case the new arrangement provides very convenient 20-digit shifts, as well as certain other manipulations. These things will appear in more detail in the discussion immediately preceding the static coding in 10.7.

10.0 CODING OF SOME SIMPLE ANALYTICAL PROBLEMS

10.1 We will code in this chapter two problems which are typical constituents of analytical problems, and in which the approximation character of numerical procedures is particularly emphasized. This is the process of numerical integration and the process of interpolation, both for a tabulated one-variable function. Both problems allow a considerable number of variants, and we will, of course, make no attempt to exhaust these. We will nevertheless vary the formulations somewhat, and discuss an actual total of six specific problems.

10.2 We consider first the integration problem.

We assume that an integral $\int f(z)dz$ is wanted, where the function $f(z)$ is defined in the z-interval 0, 1, and assumes values within the range of our machine, i.e., size < 1. Actually $f(z)$ will be supposed to be given at N+1 equidistant places in the interval 0, 1, i.e., the values $f(h/N)$, $h = 0, 1, \ldots, N$, are tabulated. We wish to evaluate the integral by numerical integration formulae with error terms of the order of those in Simpson's formula, i.e., $O(1/N^4)$.

This is a method to derive such formulae. Consider the expression

$$(1) \qquad \varphi(u) = \int_{\frac{h-\xi u}{N}}^{\frac{h+\xi u}{N}} f(z)dz - \left[2f\left(\frac{h}{N}\right)\xi + \frac{1}{3}\left(f\left(\frac{h+u}{N}\right) - 2f\left(\frac{h}{N}\right) + f\left(\frac{h-u}{N}\right)\right)\xi^3\right]\frac{u}{N} .$$

There will be $h = 0, 1, \ldots, N$, $0 \leqq \xi \leqq 1$, $0 \leqq u \leqq 1$.

Simple calculations, which we need not give here explicitly, give $\varphi(0) = \varphi'(0) = \varphi''(0) = 0$ (indeed $\varphi'''(0) = 0$ and for $\xi = 1$ even $\varphi''''(0) = 0$), implying

$$(2) \qquad \varphi(u) = \int_0^{u} du_1 \int_0^{u_1} du_2 \int_0^{u_2} du_3\, \varphi'''(u_3),$$

and further

$$(3) \qquad \varphi'''(u) = \left[f''\left(\frac{h+\xi u}{N}\right) + f''\left(\frac{h-\xi u}{N}\right)\right]\frac{\xi^3}{N^3} - \left[f''\left(\frac{h+u}{N}\right) + f''\left(\frac{h-u}{N}\right)\right]\frac{\xi^3}{N^3} -$$

$$- \frac{1}{3}\left[f'''\left(\frac{h+u}{N}\right) - f'''(h-u)\right]\frac{\xi^3 u}{N^4} .$$

(3) transforms into

$$\varphi'''(u) = [f''(\tfrac{h+\xi u}{N}) - f''(\tfrac{h+u}{N}) + f'''(\tfrac{h+u}{N})(1-\xi)\tfrac{u}{N}]\,\frac{\xi^3}{N^3} +$$

(4) $$+ [f''(\tfrac{h-\xi u}{N}) - f''(\tfrac{h-u}{N}) - f'''(\tfrac{h-u}{N})(1-\xi)\tfrac{u}{N}]\,\frac{\xi^3}{N^3} -$$

$$- [f'''(\tfrac{h+u}{N}) - f'''(\tfrac{h-u}{N})]\,(\tfrac{4}{3} - \xi)\,\frac{\xi^3 u}{N^4}$$

Putting

$$Mf'''' = \operatorname*{Max}_{z} |f''''(z)| ,$$

we see that (4) yields

(5) $$|\varphi'''(u)| \leqq 2Mf''''\,(\tfrac{7}{3} - 2\xi)\,\frac{\xi^3 u^2}{N^5} ,$$

and by (2)

(6) $$|\varphi(u)| \leqq \frac{Mf''''}{30}\,(\tfrac{7}{3} - 2\xi)\,\frac{\xi^3 u^5}{N^5} .$$

Putting u = 1, and recalling (1), we find:

(7) $$\int_{\frac{h-\xi}{N}}^{\frac{h+\xi}{N}} f(z)dz = [2f(\tfrac{h}{N})\xi + \tfrac{1}{3}(f(\tfrac{h+1}{N}) - 2f(\tfrac{h}{N}) + f(\tfrac{h-1}{N}))\,\xi^3]\,\frac{1}{N} + \varphi,$$

$$|\varphi| \leqq \frac{Mf''''}{90}\,(7-6\xi)\,\xi^3\,\frac{1}{N^5} .$$

Putting $\xi = 1$ and summing over $h = \ell+1, \ell+3, \ldots, k-1$ (ℓ, $k = 0, 1, \ldots, N$, $k-\ell>0$ and even) we get Simpson's formula:

(8) $$\int_{\frac{\ell}{N}}^{\frac{k}{N}} f(z)dz = \sum_{h=\ell+1,\ell+3,\ldots,k-1} [f(\tfrac{h+1}{N}) + 4f(\tfrac{h}{N}) + f(\tfrac{h-1}{N})]\,\frac{1}{3N} + \varphi_1 =$$

$$= \begin{bmatrix} f(\tfrac{\ell}{N}) + 4f(\tfrac{\ell+1}{N}) + 2f(\tfrac{\ell+2}{N}) + 4f(\tfrac{\ell+3}{N}) + \ldots + \\ + 4f(\tfrac{k-3}{N}) + 2f(\tfrac{k-2}{N}) + 4f(\tfrac{k-1}{N}) + f(\tfrac{k}{N}) \end{bmatrix} \frac{1}{3N} + \varphi_1 ,$$

$$|\varphi_1| \leqq \frac{Mf''''}{90}\,\frac{k-\ell}{2N^5} \leqq \frac{Mf''''}{180}\,\frac{1}{N^4} .$$

Putting $\xi = \frac{1}{2}$ and summing over $h = \ell+1, \ell+2, \ldots, k-1$ (ℓ, $k = 0, 1, \ldots, N$, $k-\ell>0$ and of any parity) we get the related formula:

$$\int_{\frac{\ell+\frac{1}{2}}{N}}^{\frac{k-\frac{1}{2}}{N}} f(z)dz = \sum_{h=\ell+1,\ell+2,\ldots,k-1} \left[f\left(\frac{h+1}{N}\right) + 22f\left(\frac{h}{N}\right) + f\left(\frac{h-1}{N}\right)\right] \frac{1}{24N} + \varphi_2 =$$

$$(9) \qquad = \left[f\left(\frac{\ell+1}{N}\right) + f\left(\frac{\ell+2}{N}\right) + \ldots + f\left(\frac{k-1}{N}\right)\right] \frac{1}{N} +$$

$$+ \left[f\left(\frac{\ell}{N}\right) - f\left(\frac{\ell+1}{N}\right) - f\left(\frac{k-1}{N}\right) + f\left(\frac{k}{N}\right)\right] \frac{1}{24N} + \varphi_2 ,$$

$$|\varphi_2| \leq \frac{Mf''''}{180} \frac{k-\ell-1}{N^5} \leq \frac{Mf''''}{180} \frac{1}{N^4} .$$

10.3 We evaluate first by Simpson's formula, i.e., by (8), and in order to simplify matters we put $\ell = 0$, $k = N$. Hence N must be even in this case

PROBLEM 10:

The function $f(z)$, z in 0, 1, is given to the extent that the values $f(h/N)$, $h = 0, 1, \ldots, N$, are stored at $N + 1$ consecutive memory locations $p, p+1, \ldots, p+N$. It is desired to evaluate the integral $\int_0^1 f(z)dz$ by means of the formula (8). ----

We could use either form of (8)'s right side for this evaluation. Using the first form leads to an induction over the odd integers $h = 1, 3, \ldots, N-1$, but requires at each step the preliminary calculation of the expression $[(f(h+1)/N) + 4f(h/N) + f((h-1)/N)]/3N$. Using the second form leads to an induction over the integers $h = 0, 1, \ldots, N$, and requires at each step the simpler expression $(2/3N)[f(h/N)]$ multiplied by a factor $\epsilon_h = \frac{1}{2}$ or 1 or 2. This factor is best handled by variable remote connections. A detailed comparison shows that the first method is about 20% more efficient, both regarding the space required by the coded sequence and the time consumed by the actual calculation. We will nevertheless code this Problem according to the second method, because this offers a good opportunity to exemplify the use of variable remote connections.

Thus the expression to be computed is

$$J = \sum_{h=0}^{N} \epsilon_h \frac{2}{3N} f\left(\frac{h}{N}\right) ,$$

$$\epsilon_h \begin{cases} = \frac{1}{2} & \text{for } h = 0, N \\ = 1 & \text{for } h \neq 0, N \text{ and } h \text{ even} \\ = 2 & \text{for } h \neq 0, N \text{ and } h \text{ odd} \end{cases} .$$

This amounts to the inductive definition

$$J_{-1} = 0,$$

$$J_h = J_{h-1} + \epsilon_h \frac{2}{3N} f(\frac{h}{N}) \quad \text{for } h = 0, 1, \ldots, N,$$

$$' = J_N$$

Let A be the storage area corresponding to the interval of locations from p to p + N. In this way its positions will be A.0, 1,...,N, where A.h corresponds to p + h, and stores f(h/N). These positions, however, are supposed to be parts of some other routine, already in the machine, and therefore they need not appear when we pass (at the end of the whole procedure) to the final enumeration of the coded sequence. (Cf. the analogous situation in Problem 3.)

Let B be the storage area which holds the given data (the constants) of the problem p, N. (It will actually be convenient to store p, N as p_0, $(p+N-1)_0$). Storage will also have to be provided for various other quantities ($0, 1_0$; it is convenient to store 2/3N; the exit-locations of the variable remote connections), these too will be accomodated in B. Next, the induction index h must be stored. h is really relevant in the combination p + h, which is a position mark (for A.h, storing f(h/N)), so we will store $(p+h)_0$ instead. (Cf. the analogous situation in Problem 3.) This will be stored at C. Finally the quantities which are processed during each inductive step will be stored in the storage area D.

These things being understood, we can now draw the flow diagram, as shown in Figure 10.1. It will prove to be convenient to store (2/3N) f(h/N) after II as well as ϵ_h (2/3N) f(h/N) after III and IV (but not after V) not at D.2, but in the accumulator, and to transfer it to D.2 in VI only (but also in V). The disposal of α_1, α_2, α_3, shown in IX, VIII, VI will be delayed until X, and they will be held over in the accumulator. Finally, the transfer of $(A.h+1)_0$ into C is better delayed from X until after the Cc order in XI, whereby it takes place only on the - branch issuing from XI, but this is the only branch on which it is needed.

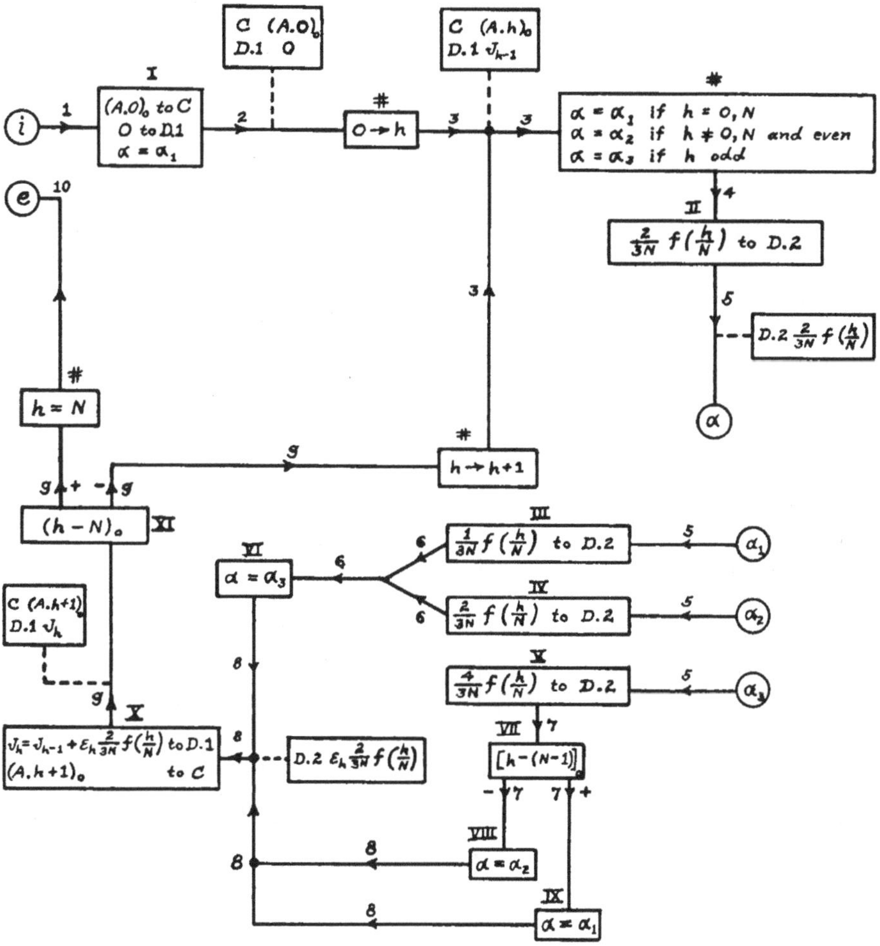

A.h	$f(\frac{h}{N})$

p, N

h

Figure 10.1

The static coding of the boxes I-XI follows:

B.1	0				
I,1	B.1		Ac	0	
2	D.1	S	D.1	0	
B.2	p_0				
I,3	B.2		Ac	p_0	
4	C	S	C	p_0	
B.3	$(\alpha_1)_0$				
I,5	B.3		Ac	$(\alpha_1)_0$	
6	II,5	Sp	II,5	α_1	C
(to II,1)					
II,1	C		Ac	$(p+h)_0$	
2	II,3	Sp	II,3	$p+h$	R
3	-	R			
[	p+h	R]	R	$f(\frac{h}{N})$	
B.4	$\frac{2}{3N}$				
II,4	B.4	^	Ac	$\frac{2}{3N} f(\frac{h}{N})$	
5	-	C			
[	α	C]			
III,1		R	Ac	$\varepsilon_h \frac{2}{3N} f(\frac{h}{N}) = \frac{1}{3N} f(\frac{h}{N})$	
(to VI,1)					
IV					
(to VI,1)					
V,1		L	Ac	$\varepsilon_h \frac{2}{3N} f(\frac{h}{N}) = \frac{4}{3N} f(\frac{h}{N})$	
2	D.2	S	D.2	$\varepsilon_h \frac{2}{3N} f(\frac{h}{N})$	
(to VII,1)					
VI,1	D.2	S	D.2	$\varepsilon_h \frac{2}{3N} f(\frac{h}{N})$	
B.5	$(\alpha_3)_0$				
VI,2	B.5		Ac	$(\alpha_3)_0$	
(to X,1)					
VII,1	C		Ac	$(p+h)_0$	
B.6	$(p+N-1)_0$				
VII,2	B.6	h-	Ac	$(h-(N-1))_0$	
3	IX,1	Cc			
(to VIII,1)					
B.7	$(\alpha_2)_0$				
VIII,1	B.7		Ac	$(\alpha_2)_0$	
(to X,1)					
IX,1	B.3		Ac	$(\alpha_1)_0$	
(to X, 1)					

X,1	II,5	Sp	II,5	α	C
2	D.1		Ac	J_{h-1}	
3	D.2	h	Ac	$J_h = J_{h-1} + \epsilon_h \frac{2}{3N} f(\frac{h}{N})$	
4	D.1	S	D.1	J_h	
(to XI,1)					
XI,1	C		Ac	$(p+h)_0$	
2	B.6	h-	Ac	$(h-(N-1))_0$	
B.8	1_0				
XI,3	B.8	h-	Ac	$(h-N)_0$	
4	e	Cc			
5	C		Ac	$(p+h)_0$	
6	B.8	h	Ac	$(p+h+1)_0$	
7	C	S	C	$(p+h+1)_0$	
(to II,1)					

Note, that box IV required no coding, hence its immediate successor (VI,1) must follow directly upon its immediate predecessor. However, this box has actually no immediate predecessor; IV,1 corresponds to α_2. Hence it must be replaced there by VI,1.

The ordering of the boxes is I, II; III, VI, X, XI; V, VII, VIII; IX (IV omitted, cf. above) and X, X, II must also be the immediate successors of VIII, IX, XI, respectively. This necessitates the extra orders

VIII,2	X,1	C
IX,2	X,1	C
XI,8	II,1	C

α_1, α_2, α_3, correspond to III,1, VI,1 (instead of IV,1), V,1. In the final enumeration the three α's must obtain numbers of the same parity. This may necessitate the insertion of dummy (ineffective, irrelevant) orders in appropriate places, which we will mark *.

We must now assign B.1-8, C, D.1-2 their actual values, pair the 35 orders I,1-6, II,1-5, III,1, V,1-2, VI,1-2, VII,1-3, VIII,1-2, IX,1-2, X,1-4, XI,1-8 to 18 words, and then assign I,1-XI,8 their actual values. These are expressed in this table:

I,1-6	0 - 2'	X,1-4	7' - 9	VIII,1-2	16 - 16'
II,1-5	3 - 5	XI,1-8	9' - 13	IX,1-2	17 - 17'
III,1,*	5' - 6	V,1-2	13' - 14	B.1-8	18 - 25
VI,1-2	6' - 7	VII,1-3	14' - 15'	C	26
				D.1-2	27 - 28

Now we obtain this coded sequence:

0	18 , 27 S	10	23 h-, 25 h-	20	5_o
1	19 , 26 S	11	e Cc, 26	21	2/3N
2	20 , 5 Sp	12	25 h , 26 S	22	13_o
3	26 , 4 Sp	13	3 C , L	23	$(p+N-1)_o$
4	- R , 21 x	14	28 S , 26	24	6_o
5	- C', R	15	23 h-, 17 Cc	25	1_o
6	-- , 28 S	16	24 , 7 C'	26	--
7	22 , 5 Sp	17	20 , 7 C'	27	--
8	27 , 28 h	18	0	28	--
9	27 S , 26	19	p_o		

The durations may be estimated as follows:

I: 225 μ, II: 270 μ, III: 30 μ, V: 55 μ, VI: 75 μ: VII: 125 μ, VIII: 75 μ, IX: 75 μ, X: 150 μ, XI: 300 μ.

Total: $\mathrm{I} + \mathrm{II} \times (N+1) + \mathrm{III} \times 2 + \mathrm{VI} \times (\frac{N}{2}+1) + (\mathrm{V} + \mathrm{VII}) \times \frac{N}{2} +$

$+ \mathrm{VIII} \times (\frac{N}{2} - 1) + \mathrm{IX} + (\mathrm{X} + \mathrm{XI}) \times (N+1) =$

$= (225 + 270\,(N+1) + 60 + 75\,(\frac{N}{2}+1) + 180\,\frac{N}{2} + 75\,(\frac{N}{2} - 1) + 75 + 450\,(N+1))\,\mu =$

$= (885\,N + 1080)\,\mu \approx (.9\,N + 1.1)\,m.$

10.4 We evaluate next by the formula (9), and this time we keep ℓ, k general. We had ℓ, k = 0, 1,...,N and $k-\ell>0$. This excludes k = 0 as well as ℓ = N. It is somewhat more convenient to write ℓ-1 for ℓ. Then ℓ, k = ℓ,...,N and $k-\ell\geq 0$. Thus (9) becomes

$$(10)\qquad \int_{\frac{\ell-\frac12}{N}}^{\frac{k-\frac12}{N}} f(z)dz = [f(\tfrac{\ell}{N}) + f(\tfrac{\ell+1}{N}) + \ldots + f(\tfrac{k-1}{N})]\,\frac{1}{N} + [f(\tfrac{\ell-1}{N}) - f(\tfrac{\ell}{N}) - f(\tfrac{k-1}{N}) + f(\tfrac{k}{N})]\,\frac{1}{24N} + \varphi_2\,,$$

$$|\varphi_2| \leq \frac{Mf''''}{180}\,\frac{k-\ell}{N^5} \leq \frac{Mf''''}{180}\,\frac{1}{N^4}\,.$$

Finally the requirement $k-\ell\geq 0$ may be dropped, since for $k-\ell\leq 0$

$$\int_{\frac{\ell-\frac12}{N}}^{\frac{k-\frac12}{N}} = -\int_{\frac{k-\frac12}{N}}^{\frac{\ell-\frac12}{N}}\,.$$

Thus N and ℓ, k = 1,...,N are subject to no further restrictions.

We state accordingly:

PROBLEM 11.

Same as Problem 10, with this change. It is desired to evaluate the integral $\int_{\frac{\ell-\frac{1}{2}}{N}}^{\frac{k-\frac{1}{2}}{N}} f(z)dz$ (ℓ, k = 1,...,N) by means of the formula (10) (for $k \geq \ell$, for $k < \ell$ interchange ℓ, k). ----

The expression to be computed is

$$J = \pm J' \text{ for } \begin{Bmatrix} k \geq \ell \\ k < \ell \end{Bmatrix}$$

where J' is defined as follows: Put

$$k' = \begin{Bmatrix} k \\ \ell \end{Bmatrix}, \quad \ell' = \begin{Bmatrix} \ell \\ k \end{Bmatrix} \quad \text{for} \begin{Bmatrix} k \geq \ell \\ k < \ell \end{Bmatrix},$$

then

$$J' = J'' + \frac{1}{24N} \left[f\left(\frac{\ell'-1}{N}\right) - f\left(\frac{\ell'}{N}\right) - f\left(\frac{k'-1}{N}\right) + f\left(\frac{k'}{N}\right) \right],$$

and

$$J'' = \frac{1}{N} \sum_{h=\ell'}^{k'-1} f\left(\frac{h}{N}\right)$$

The last equation amounts to the inductive definition

$$J_{\ell'-1} = 0,$$

$$J_h = J_{h-1} + \frac{1}{N} f\left(\frac{h}{N}\right),$$

$$J'' = J_{k'-1}$$

The storage areas A, B, C, D and the induction index h will be treated the same way as in Problem 10, but h runs now over ℓ', $\ell'+1, \ldots, k'-1$.

We can now draw the flow diagram as shown in Figure 10.2. It will be found convenient to store the contents of E.1-2, $(p+k')_0$, $(p+\ell')_0$, in the same place where $(p+k)_0$, $(p+\ell)_0$ are originally stored, which proves to be B.1-2. This simplifies II and reduces the memory requirements, but since we wish to have B.1-2 at the end of the routine in the same state in which they were at the beginning, it requires restoring B.1-2 in IX. VIII, on the other hand, turns out to be entirely unnecessary.

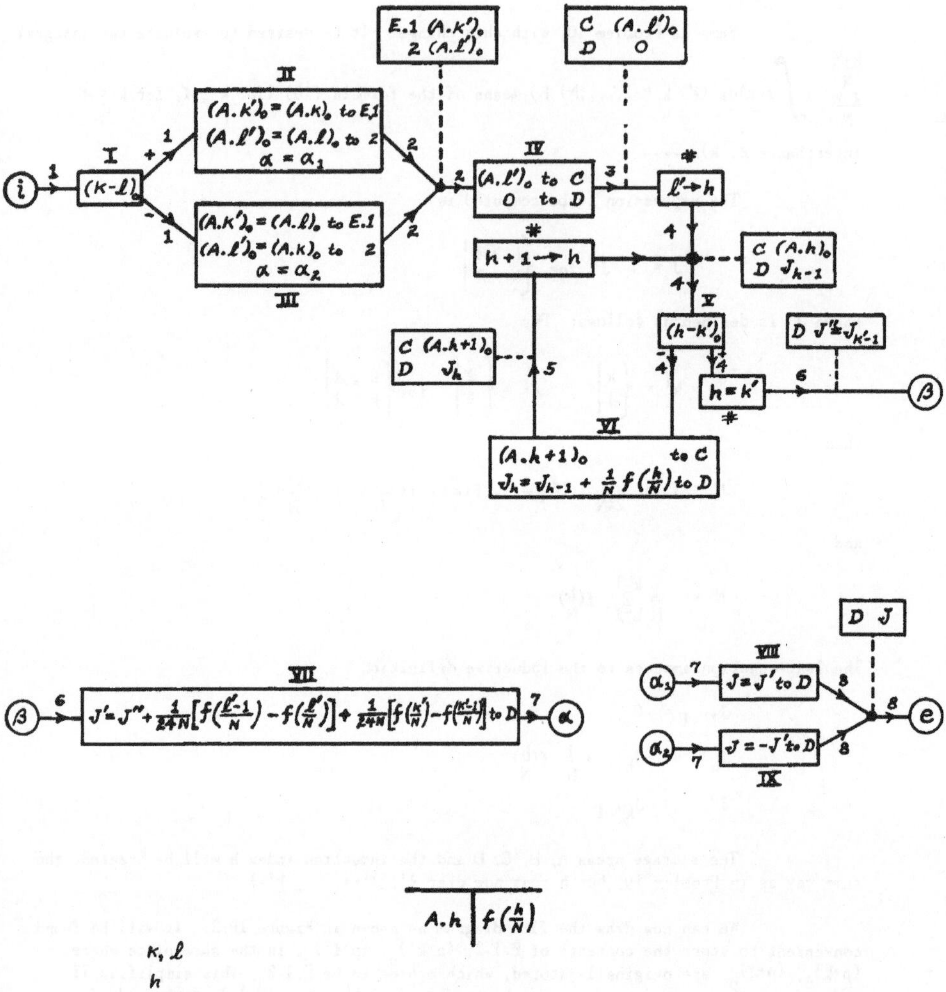
E.1 (A.K')$_0$
2 (A.ℓ')$_0$
C (A.ℓ')$_0$
D 0
II
(A.K')$_0$ = (A.K)$_0$ to E.1
(A.ℓ')$_0$ = (A.ℓ)$_0$ to 2
α = α$_1$
I
(K-ℓ)$_0$
(A.K')$_0$ = (A.ℓ)$_0$ to E.1
(A.ℓ')$_0$ = (A.K)$_0$ to 2
α = α$_2$
III
IV
(A.ℓ')$_0$ to C
0 to D
ℓ' → h
h + 1 → h
C (A.h)$_0$
D J$_{h-1}$
V
(h-K')$_0$
D J'' = J$_{K'-1}$
C (A.h+1)$_0$
D J$_h$
h = K'
β
VI
(A.h+1)$_0$ to C
J$_h$ = J$_{h-1}$ + $\frac{1}{N}$ f($\frac{h}{N}$) to D
D J
VII
J' = J'' + $\frac{1}{24N}$[f($\frac{ℓ'-1}{N}$) - f($\frac{ℓ'}{N}$)] + $\frac{1}{24N}$[f($\frac{K'}{N}$) - f($\frac{K'-1}{N}$)] to D
VIII
J = J' to D
IX
J = -J' to D
A.h | f($\frac{h}{N}$)
K, ℓ
h

FIGURE 10.2

The static coding of the boxes I-IX follows:

B.1	$(p+k)_0$				
2	$(p+\ell)_0$.				
I,1	B.1		Ac	$(p+k)_0$	
2	B.2	h-	Ac	$(k-\ell)_0$	
3	II,1	Cc			
(to III,1)					
B.3	$(\alpha_1)_0$				
II,1	B.3		Ac	$(\alpha_1)_0$	
2	VII,23	Sp	VII,23	α_1	C
(to IV,1)					
III,1	B.1		Ac	$(p+k)_0$	
2	s.1	S	s.1	$(p+k)_0$	
3	B.2		Ac	$(p+\ell)_0$	
	B.1	S	B.1	$(p+\ell)_0$	
5	s.1		Ac	$(p+k)_0$	
6	B.2	S	B.2	$(p+k)_0$	
B.4	$(\alpha_2)_0$				
III,7	B.4		Ac	$(\alpha_2)_0$	
8	VII,23	Sp	VII,23	α_2	C
(to IV,1)					
IV,1	B.2		Ac	$(p+\ell')_0$	
2	C	S	C	$(p+\ell')_0$	
B.5	0				
IV,3	B.5		Ac	0	
4	D	S	D	0	
(to V,1)					
V,1	C		Ac	$(p+h)_0$	
2	B.1	h-	Ac	$(h-k')_0$	
3	VII,1	Cc			
(to VI,1)					
VI,1	C		Ac	$(p+h)_0$	
2	VI,3	Sp	VI,3	p+h	R
3	-	R			
[	p+h	R]	R	$f(\frac{h}{N})$	
B.6	$\frac{1}{N}$				
VI,4	B.6	x	Ac	$\frac{1}{N} f(\frac{h}{N})$	
5	D	h	Ac	$J_h = J_{h-1} + \frac{1}{N} f(\frac{h}{N})$	
6	D	S	D	J_h	
7	C		Ac	$(p+h)_0$	
B.7	1_0				

VI,8	B.7	h	Ac	$(p+h+1)_0$	
9	C	S	C	$(p+h+1)_0$	
(to V,1)					
VII,1	B.2		Ac	$(p+\chi')_0$	
2	VII,6	Sp	VII,6	$p+\chi'$	h-
3	B.7	h-	Ac	$(p+\chi'-1)_0$	
4	VII,5	Sp	VII,5	$p+\chi'-1$	
5	--				
[	$p+\chi'-1$	]	Ac	$f(\frac{\chi'-1}{N})$	
6	--	h-			
[	$p+\chi'$	h-]	Ac	$f(\frac{\chi'-1}{N}) - f(\frac{\chi'}{N})$	
	s.1	S	s.1	$f(\frac{\chi'-1}{N}) - f(\frac{\chi'}{N})$	
8	s.1	R	R	$f(\frac{\chi'-1}{N}) - f(\frac{\chi'}{N})$	
B.8	$\frac{1}{24N}$				
VII,9	B.8		Ac	$\frac{1}{24N}\ (f(\frac{\chi'-1}{N}) - f(\frac{\chi'}{N}))$	
10	D	h	Ac	$J'' + \frac{1}{24N}\ (f(\frac{\chi'-1}{N}) - f(\frac{\chi'}{N}))$	
11	D	S	D	$J'' + \frac{1}{24N}\ (f(\frac{\chi'-1}{N}) - f(\frac{\chi'}{N}))$	
12	B.1		Ac	$(p+k')_0$	
13	VII,17	Sp	VII,17	$p+k'$	h
14	B.7	h-	Ac	$(p+k'-1)_0$	
15	VII,16	Sp	VII,16	$p+k'-1$	
16	--	-			
[	$p+k'-1$	-]	Ac	$-f(\frac{k'-1}{N})$	
17	--	h			
[	$p+k'$	h]	Ac	$f(\frac{k'}{N}) - f(\frac{k'-1}{N})$	
18	s.1	S	s.1	$f(\frac{k'}{N}) - f(\frac{k'-1}{N})$	
19	s.1	R	R	$f(\frac{k'}{N}) - f(\frac{k'-1}{N})$	
20	B.8		Ac	$\frac{1}{24N}\ (f(\frac{k'}{N}) - f(\frac{k'-1}{N}))$	
21	D	h	Ac	$J' = J'' + \frac{1}{24N}\ (f(\frac{\chi'-1}{N}) - f(\frac{\chi'}{N})) + \frac{1}{24N}\ (f(\frac{k'}{N}) - f(\frac{k'-1}{N}))$	
22	D	S	D	J'	
23	--	C			
[	α	C]			

VIII				
(to e)				
IX,1	D	-	Ac	$J = -J'$
2	D	S	D	J
3	B.1		Ac	$(p+\ell)_0 = (p+k')_0$
4	s.1	S	s.1	$(p+\ell)_0$
5	B.2		Ac	$(p+k)_0 = (p+\ell')_0$
6	B.1	S	B.1	$(p+k)_0$
7	s.1		Ac	$(p+\ell)_0$
8	B.2	S	B.2	$(p+\ell)_0$
(to e)				

Note, that the box VIII required no coding, hence its immediate successor (e) must follow directly upon its immediate predecessor. However, this box has actually no immediate predecessor; VIII,1 corresponds to α_1, which may appear (by substitution at II,2) in the C order VII,23. Hence VIII,1 must be replaced by e in α_1.

The ordering of the boxes is I, III, IV, V, VI; II; VII; IX (VIII omitted, cf. above), and IV, V, e must also be the immediate successors of II, VI, IX, respectively. This necessitates the extra orders:

II,3	IV,1	C
VI,10	V,1	C
IX,9	e	C

α_1 corresponds to VIII,1, i.e., to e (cf. above), α_2 corresponds to IX,1. This implies, as in the corresponding situation in Problem 10, that IX,1 and e must have in the final enumeration numbers of the same parity. β need not be considered, since it represents a fixed remote connection and therefore does not appear in the above detailed coding.

We must now assign B.1-8, C, D, s.1 their actual values, pair the 63 orders I,1-3, II,1-3, III,1-8, IV,1-4, V,1-3, VI,1-10, VII,1-23, IX,1-9 to 32 words, and then assign I,1-IX,9 their actual values. These are expressed in this table:

I,1-3	0 - 1	VI,1-10	9 - 13'	B.1-8	32 - 39
III,1-8	1'- 5	II,1-3	14 - 15	C	40
IV,1-4	5'- 7	VII,1-23	15'- 26'	D	41
V,1-3	7'- 8'	IX,1-9	27 - 31	s.1	42

e is supposed to have the parity of IX,1, i.e., to be unprimed.

Now we obtain this coded sequence:

0	32 ,	33 h-	14	34 ,	26 Sp'	28	32 ,	42 S
1	14 Cc,	32	15	5 C',	33	29	33 ,	32 S
2	42 S ,	33	16	18 Sp,	38 h-	30	42 ,	33 S
3	32 S ,	42	17	17 Sp',	-	31	e C ,	--
4	33 S ,	35	18	- h-,	42 S	32	$(p+k)_0$	
5	26 Sp',	33	19	42 R ,	39 x	33	$(p+\ell)_0$	
6	40 S ,	36	20	41 h ,	41 S	34	e_0	
7	41 S ,	40	21	32 ,	23 Sp'	35	27_0	
8	32 h-,	15 Cc'	22	38 h-,	23 Sp	36	0	
9	40 ,	10 Sp	23	- - ,	- h	37	$\frac{1}{N}$	
10	- R,	37 x	24	42 S ,	42 R	38	1_0	
11	41 h ,	41 S	25	39 x ,	41 h	39	$\frac{1}{24N}$	
12	40 ,	38 h	26	41 S ,	- C	40	--	
13	40 S ,	7 C'	27	41 - ,	41 S	41	--	
						42	--	

The durations may be estimated as follows:

I: 125 μ, II: 125 μ, III: 300 μ, IV: 150 μ, V: 125 μ, VI: 445 μ, VII: 1015 μ, IX: 350 μ.

Total: $\text{I} + (\text{II or III}) + \text{IV} + \text{V} \times (|k-\ell| + 1) + \text{VI} \times |k-\ell| + \text{VII} + (\text{VIII or IX})$

$$\text{maximum} = (125 + 300 + 150 + 125\,(|k-\ell| + 1) + 445\,|k-\ell| + 1{,}015 + 350\ \mu =$$

$$= (570\,|k-\ell| + 2{,}065)\ \mu$$

$$\text{maximum} = (570\,(N-1) + 2{,}065)\ \mu = (570\,N + 1{,}495)\ \mu \approx (.6\,N + 1.5)\ \text{m}.$$

10.5 We pass now to the interpolation problem.

Lagrange's interpolation formula expresses the unique polynomial $P(x)$ of degree $M-1$, which assumes M given values $p_1,\dots,p_M$ at M given places $x_1,\dots,x_M$, respectively:

$$(1)\qquad P(x) = P(x_1, p_1;\dots; x_M, p_M \,|\, x) = \sum_{i=1}^{M} p_i \frac{\prod_{j=1}^{M(j\neq i)} (x - x_j)}{\prod_{j=1}^{M(j\neq i)} (x_i - x_j)}$$

There would be no difficulty in devising a (multiply inductive) routine which evaluates the right hand side of (1) directly. This, however, seems inadvisable, except for relatively small values of M. The reason is that the denominators $\prod_{j=1}^{M(j\neq i)}$ are likely to be inadmissibly small.

This point may deserve a somewhat more detailed analysis.

From our general conditions of storage and the speed of our arithmetical organs one will be inclined to conclude that the space allotted to the storage of the functions which are evaluated by interpolation should (in a given problem) be comparable to the space occupied by the interpolation routine itself. The latter amounts to about 100 words. (Problem 12 occupies together with Problem 13.a or 13.b or 13.c 99 or 101 or 106 words, respectively, cf. 10.7 or 10.8 or 10.9, respectively. Other possible variants occupy very comparable amounts of space.) One problem will frequently use interpolation on several functions. It seems therefore reasonable to expect that each of these functions will be given at $\sim \frac{1}{3} \cdot 100 \sim 2^5$ points. (This means that the N of 10.7 will be $\sim 2^5$ -- not our present M! Note also, that the storage required in this connection is N in Problem 13.a, but 2N in Problems 13.b and 13.c.) Hence we may expect that the points $x_1, \ldots, x_M$ will be at distances of the order $\sim 2^{-5}$ between neighbors.

Hence $|x_i - x_j| \sim |i-j| \cdot 2^{-5}$, and so

$$\left| \prod_{j=1}^{M (j \neq i)} (x_i - x_j) \right| \sim (i-1)!\,(M-i)!\,2^{-5(M-1)} .$$

The round-off errors of our multiplication introduce into all these products absolute errors of the order 2^{-40}. Hence the denominators $\prod_{j=1}^{M (j \neq i)} (x_i - x_j)$, and with them the corresponding terms of the sum $\sum_{i=1}^{M}$ in (1), are affected with relative errors of the order $2^{-40} : (i-1)!\,(M-i)!\,2^{-5(M-1)} = \frac{2^{5M-45}}{(M-1)!} \binom{M-1}{i-1}$. The average affect of these relative errors is best estimated as the relative error

$$\sqrt{\frac{1}{M} \sum_{i=1}^{M} \left[\frac{2^{5M-45}}{(M-1)!} \binom{M-1}{i-1}\right]^2} = \frac{2^{5M-45}}{(M-1)!\,\sqrt{M}} \sqrt{\sum_{i=1}^{M} \binom{M-1}{i-1}^2} =$$

$$= \frac{2^{5M-45}}{(M-1)!\,\sqrt{M}} \sqrt{\binom{2M-2}{M-1}} = \frac{2^{5M-45}}{(M-1)!\,\sqrt{M}} \sqrt{\frac{(2(M-1))!}{[(M-1)!]^2}} =$$

$$= \sqrt{\frac{(2(M-1))!}{M}} \; \frac{1}{[(M-1)!]^2} \; 2^{5M-45} .$$

On the other hand, an interpolation of degree M-1, with an interval length $\sim 2^{-5}$, is likely to have a relative precision of the order $\sim C \cdot 2^{-5M}$, where C is a moderately large number. (The function that is being interpolated is assumed to be reasonably smooth.)

Consequently the optimum relative precision ε which can be obtained with this procedure, and the optimum M that goes with it, are determined by these conditions:

$$\varepsilon \sim \sqrt{\frac{(2(M-1))!}{M}} \quad \frac{1}{[(M-1)!]} \quad 2^{5M-45} \sim C \cdot 2^{-5M} .$$

From this

$$f_M = \sqrt{\frac{(2(M-1))!}{M}} \cdot \frac{1}{[(M-1)!]^2} \cdot 2^{10M-45} \sim C.$$

Now we have

M	3	4	5	6	7
f_M	$8\cdot 10^{-6}$	10^{-2}	5	$1.8\cdot 10^{3}$	$6\cdot 10^{5}$

The plausible values of C are in the neighborhood of M = 5, while M = 6 is somewhat high and M = 4 and 7 are extremely low and high, respectively. Hence under these conditions M = 5 (biquadratic interpolation) would seem to be normally optimal, with M = 6 a somewhat less probable possibility. We have

M	5	6
ε	$1.4\cdot 10^{-7}$	$1.7\cdot 10^{-6}$

It follows, therefore, that we may expect to obtain by a reasonable application of this method relative precisions of the order $\sim 10^{-6} \sim 2^{-20}$. This is, however, only the relative precision as delimited by one particular source of errors: The arithmetical (round-off) errors of an interpolation. The ultimate level of precision of the entire problem, in which this interpolation occurs, is therefore likely to be a good deal less favorable.

These things being understood, it seems likely that the resulting level of precision will be acceptable in many classes of problems, especially among those which originate in physics. There are, on the other hand, numerous and important problems where this is not desirable or acceptable, especially since it represents the loss of half the intrinsic precision of the 40 (binary) digit machine. It is therefore worthwhile to look for alternative procedures.

The obvious method to avoid the loss of (relative) precision caused in the formula (1) by the smallness of a denominator $\prod_{j=1}^{M(j\neq i)} (x_i - x_j)$, is to divide by its factors $x_i - x_j$ ($j = 1,\ldots,M$ and $j \neq i$) singly and successively. This must, however, be combined with a similar treatment of the numerators $\prod_{j=1}^{M(j\neq i)} (x-x_j)$, since they may cause a comparable loss of (relative) precision by the same mechanism. A possible alternative to (1), which eliminates both sources of error in the sense indicated, is

$$(2) \qquad P(x) = \sum_{i=1}^{M} p_i \prod_{j=1}^{M(j\neq i)} \frac{x-x_j}{x_i-x_j}$$

(2) involves considerably more divisions than (1), but this need not be the dominant consideration. There exists, however, a third procedure, which has all the advantages of (2), and is somewhat more easily handled. Besides, its storage and induction problems are more instructive than those of (1) or (2), and for these reasons we propose to use this third procedure as the basis of our discussion.

This procedure is based on A. C. Aitken's identity

$$(3) \qquad P(x_1, p_1; \dots; x_M, p_M \mid x) = \frac{x-x_1}{x_M-x_1} P(x_2, p_2; \dots; x_M, p_M \mid x) + \frac{x_M-x}{x_M-x_1} P(x_1, p_1; \dots; x_{M-1}, p_{M-1} \mid x) .$$

Since (3) replaces an M-point interpolation by two M-1 point interpolations, it is clearly a possible basis for an inductive procedure. It might seem, however, that the reduction from an M-point interpolation to one-point ones (i.e. to constants) will involve 2^{M-1} steps, which would be excessive, since (2) is clearly an M(M-1) step process. However, (3) removes either extremity (x_1 or x_M) of the point system $x_1, \dots, x_M$; hence iterating it can only lead to point systems $x_i, \dots, x_j$ ($i, j = 1, \dots, M$, $i \leq j$, we will write $j = i+h-1$), of which there are only $\frac{M(M+1)}{2}$; i.e., $\frac{M(M-1)}{2}$, not counting the one-point systems. Hence (3) is likely to lead to something like an $\frac{M(M-1)}{2}$ step process.

Regarding the sizes we assume that $x_1, \dots, x_M$ as well as x and $p_1, \dots, p_M$ lie in the interval -1, 1. We need, furthermore, that the differences $x_{i+h}-x_i$, $x-x_i$, $p_{i+h}-p_i$ also lie in the interval -1, 1, and it is even necessary in view of the particular algebraical routine that we use, to have all differences $p_{i+h}-p_i$ (absolutely) smaller than the corresponding $x_{i+h}-x_i$. (Cf. VII,19.) I.e., we must use appropriate size adjusting factors for the p_i to secure this "Lipschitz condition". The same must be postulated for the differences $P_{i+1}^h(x)-P_i^h(x)$ of the intermediate interpolation polynomials. All of this might be circumvented to various extents in various ways, but this would lead us deeply into the problems of polynomial interpolation which are not our concern here.

We now state:

PROBLEM 12.

The variable values $x_1,\ldots,x_M$ $(x_1 < x_2 < \ldots < x_m)$ and the function values $p_1,\ldots,p_M$ are stored at two systems of M consecutive memory locations each: q, q+1, ..., q+M-1 and p, p+1, ..., p+M-1. The constants of the problem are p, q, M, x, and they are stored at four given memory locations. It is desired to interpolate this function for the value x of the variable, using Lagrange's formula. The process of reduction based on the identity (3) is to be used. ----

It is clear from the previous discussion, that we have to form the family of interpolants

(4) $$P_i^h(x) = P(x_i, p_i; \ldots; x_{i+h-1}, p_{i+h-1} \mid x) \quad \text{for } i = 1,\ldots,M, \quad h = 1,\ldots,M-i+1 .$$

Applying (3) to $P(x_i, p_i;\ldots; x_{i+h}, p_{i+h} \mid x)$ instead of $P(x_1, p_1;\ldots; x_M, p_M \mid x)$ gives

$$P_i^{h+1}(x) = \frac{x - x_i}{x_{i+h} - x_i} P_{i+1}^h(x) + \frac{x_{i+h} - x}{x_{i+h} - x_i} P_i^h(x) ,$$

or equivalently

(5) $$P_i^{h+1}(x) = P_i^h(x) + \frac{x - x_i}{x_{i+h} - x_i}(P_{i+1}^h(x) - P_i^h(x)) .$$

Combining this with

(6) $$P_i^1(x) = p_i$$

and

(7) $$P(x) = P(x_1, p_1;\ldots; x_M, p_M \mid x) = P_1^M(x) ,$$

we have a (doubly) inductive scheme to calculate P(x).

. Let A and B be the storage areas corresponding to the two intervals of locations from p to p+M-1 and from q to q+M-1. In this way their positions will be A.1,...,M and B.1,...,M, where A.i and B.i correspond to p+i-1 and q+i-1, and store p_i and x_i, respectively. As in the Problems 3, 10, and 11, the positions of A and B need not be shown in the final enumeration of the coded sequence.

The primary induction will clearly begin with the $p_1,\ldots,p_M$, i.e., $P_1^1(x),\ldots,P_M^1(x)$, stored at A, and obtain from these $P_1^2(x),\ldots,P_{M-1}^2(x)$; then from these the $P_1^3(x),\ldots,P_{M-2}^3(x)$; from these the $P_1^4(x),\ldots,P_{M-3}^4(x)$; etc., etc., to conclude with $P_1^M(x)$, i.e., with the desired P(x). These successive stages correspond to h = 1, 2, 3,...,M, respectively. This h is the primary induction index.

In passing from h to h+1, i.e., from $P_1^h(x), \ldots, P_{M-h+1}^h(x)$ to $P_1^{h+1}(x), \ldots, P_{M-h}^{h+1}(x)$, the $P_i^{h+1}(x)$ have to be formed successively for $i = 1, \ldots, M-h$. This i is the secondary induction index.

There is clearly need for a storage area C which holds $P_1^h(x), \ldots, P_{M-h+1}^h(x)$ at each stage h. As the transition to stage h+1 takes place, each $P_i^h(x)$ will be replaced by $P_i^{h+1}(x)$, successively for all $i = 1, \ldots, M-h$. Hence the capacity required for C is M-h+1 during the stage h, but for h = 1 the $P_i^1(x) = p_i$ are still in A, and need not appear in C. Hence the maximum capacity for C is M-1. Accordingly, let C correspond to the interval of locations from r to r+M-2. In this way its positions will be C.1,...,M-1, where C.i corresponds to r+i-1, and stores $P_i^h(x)$. The positions of the area C need not be shown in the final enumeration of the coded sequence. All that is necessary is that they be available (i.e. empty or irrelevantly occupied) when the instructions of the coded sequence are being carried out by the machine.

As stated above, the $P_i^h(x)$ occupying C.i have necessarily h > 1. To give more detail: $P_i^h(x)$ moves into the location C.i at the end of the i-th step of stage h-1, and remains there during the remainder (the M-h+1-i last steps) of stage h-1 and during the i first steps of stage h.

Further storage capacities are required as follows: The given data (the constants) of the problem, p, q, r, M, x, will be stored in the storage area D. (It will be convenient to store them as p_0, $(q-p-1)_0$, r_0, $(M-1)_0$, x.) Storage will also have to be provided for various other fixed quantities (1_0, the exit-locations of the variable remote connections), these too will be accommodated in D. Next, the two induction indices h, i, will have to be stored. As before, they are both relevant as position marks, and it will prove convenient to store in their place $(M-h)_0$, $(p+i)_0$ (i.e., $(A.i+1)_0$). These will be stored in the storage area E. Finally, the quantities which are processed during each inductive step will be stored in the storage area F.

We can now draw the flow diagram, as shown in Figure 10.3. The variable remote connections α and β are necessary, in order to make the differing treatments required for h = 1 and for h > 1 possible: In the first case the $P_i^h(x)$ are equal to p_i and come from A, in the second case they come from C (cf. above). It should be noted that in this situation our flow diagram rules impose a rather detailed showing of the contents of the storage area C.

The actual coding contains a number of minor deviations from the flow diagram, inasmuch as it is convenient to move a few operations from the box in which they are shown to an earlier box. Since we had instances of this already in earlier problems, we will not discuss it here in detail. On the other hand, some further condensations of the coding, which are possible, but deviate still further from the flow diagram, will not be considered here.

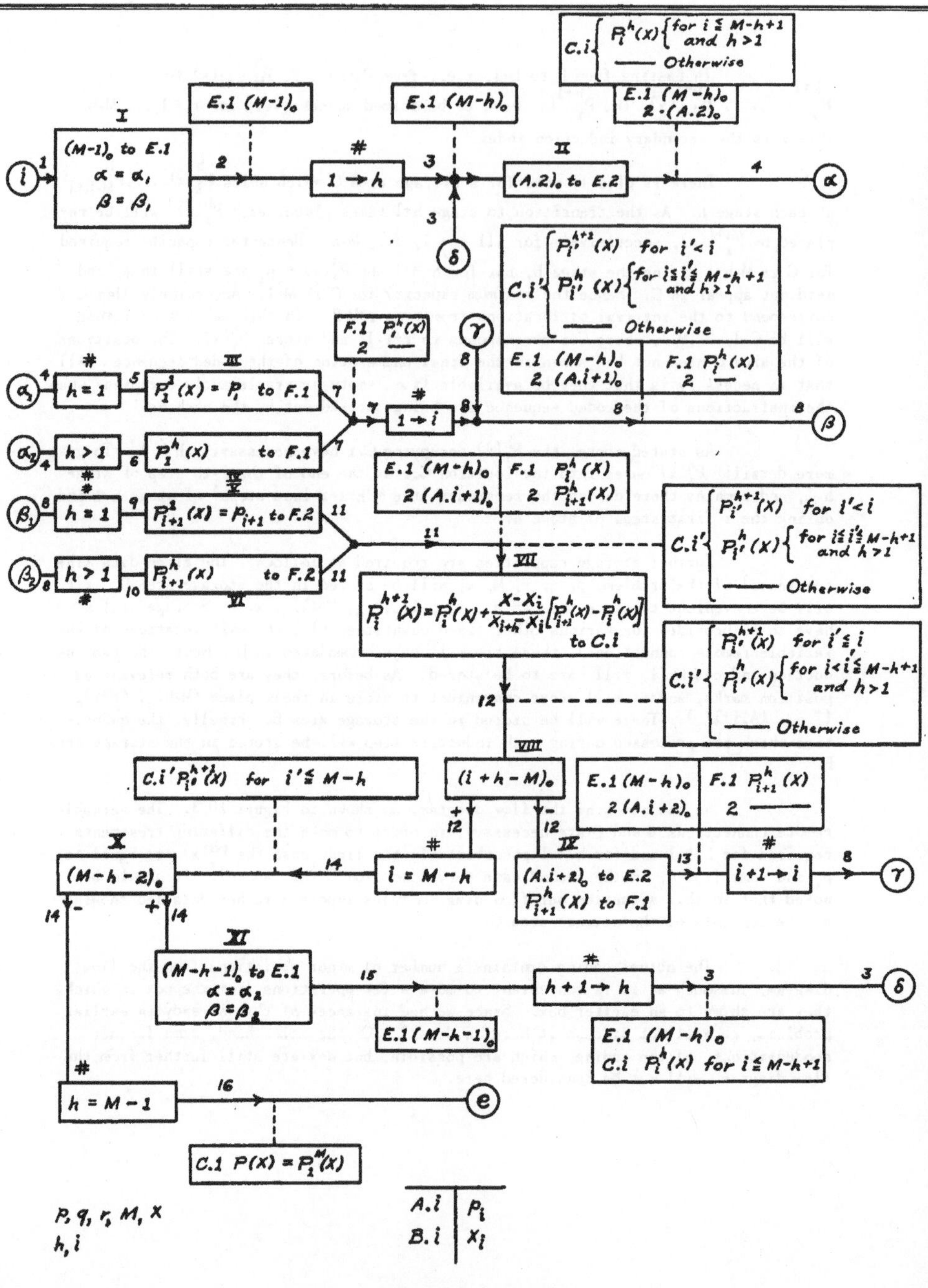

C.i Pᵢʰ(X) for i ≤ M-h+1 and h > 1; — Otherwise
E.1 (M-1)₀
E.1 (M-h)₀
E.1 (M-h)₀ 2·(A.2)₀
(M-1)₀ to E.1 α = α₁ β = β₁
1 → h
(A.2)₀ to E.2
C.i' Pᵢ'ʰ⁺¹(X) for i' < i; Pᵢ'ʰ(X) for i ≤ i' ≤ M-h+1 and h > 1; — Otherwise
F.1 Pᵢʰ(X) 2 —
E.1 (M-h)₀ 2 (A.i+1)₀
F.1 Pᵢʰ(X) 2 —
h = 1
P₁¹(X) = P₁ to F.1
h > 1
P₁ʰ(X) to F.1
1 → i
E.1 (M-h)₀ 2 (A.i+1)₀
F.1 Pᵢʰ(X) 2 Pᵢ₊₁ʰ(X)
h = 1
Pᵢ₊₁¹(X) = Pᵢ₊₁ to F.2
h > 1
Pᵢ₊₁ʰ(X) to F.2
C.i' Pᵢ'ʰ⁺¹(X) for i' < i; Pᵢ'ʰ(X) for i ≤ i' ≤ M-h+1 and h > 1; — Otherwise
Pᵢʰ⁺¹(X) = Pᵢʰ(X) + (X - Xᵢ)/(Xᵢ₊ₕ - Xᵢ)[Pᵢ₊₁ʰ(X) - Pᵢʰ(X)] to C.i
C.i' Pᵢ'ʰ⁺¹(X) for i' ≤ i; Pᵢ'ʰ(X) for i < i' ≤ M-h+1 and h > 1; — Otherwise
C.i' Pᵢ'ʰ⁺¹(X) for i' ≤ M-h
(i + h - M)₀
E.1 (M-h)₀ 2 (A.i+2)₀
F.1 Pᵢ₊₁ʰ(X) 2 —
(M-h-2)₀
i = M-h
(A.i+2)₀ to E.2 Pᵢ₊₁ʰ(X) to F.1
i+1 → i
(M-h-1)₀ to E.1 α = α₂ β = β₂
h+1 → h
E.1 (M-h-1)₀
E.1 (M-h)₀ C.i Pᵢʰ(X) for i ≤ M-h+1
h = M-1
C.1 P(X) = P₁ᴹ(X)
P, q, r, M, X
h, i
A.i Pᵢ
B.i Xᵢ

FIGURE 10.3

The static coding of the boxes I-XI follows:

	D.1	$(M-1)_0$					
	I,1	D.1			Ac	$(M-1)_0$	
	2	E.1	S		E.1	$(M-1)_0$	
	D.2	$(\alpha_1)_0$					
	I,3	D.2			Ac	$(\alpha_1)_0$	
	4	II,4	Sp		II,4	α_1	C
	D.3	$(\beta_1)_0$					
	I,5	D.3			Ac	$(\beta_1)_0$	
	6	III,6	Sp		III,6	β_1	C
	(to II,1)						
	D.4	p_0					
	5	1_0					
	II,1	D.4			Ac	p_0	
	2	D.5	h		Ac	$(p+1)_0$	
	3	E.2	S		E.2	$(p+1)_0$	
		--	C				
[		α	C	]			
	III,1	D.4			Ac	p_0	
	2	III,3	Sp		III,3	p	
	3	--					
[		p		]	Ac	$p_1^1(x) = p_1$	
		F.1	S		F.1	p_1 if reached from III,3	
						$p_1^h(x)$ if reached from IV	
	5	E.2			Ac	$(p+1)_0$ if reached from III,4	
						$(p+i)_0$ if reached from IX (i.e. γ)	
	6	--	C				
[		β	C	]			
	D.6	r_0					
	IV,1	D.6			Ac	r_0	
	2	IV,3	Sp		IV,3	r	
	3	--					
[				]	Ac	$p_1^h(x)$	
	(to III,4)						
	V,1	V,2	Sp		V,2	p+i	
	2	--					
[		p+i		]	Ac	p_{i+1}	
	3	F.2	S		F.2	p_{i+1}	
	(to VII,1)						
	VI,1	D.4	h-		Ac	i_0	
	2	D.6	h		Ac	$(r+i)_0$	
	3	VI,4	Sp		VI,4	r+i	
	4	--					
[		r+i		]	Ac	$p_{i+1}^h(x)$	
		F.2	S		F.2	$p_{i+1}^h(x)$	

(to VII,1)

VII,1	E.2		Ac	$(p+i)_o$	
2	D.4	h-	Ac	i_o	
3	D.6	h	Ac	$(r+i)_o$	
	D.5	h-	Ac	$(r+i-1)_o$	
5	VII,25	Sp	VII,25	r+i-1	S
6	E.2		Ac	$(p+i)_o$	
D.7	$(q-p-1)_o$				
VII,7	D.7	h	Ac	$(q+i-1)_o$	
8	VII,15	Sp	VII,15	q+i-1	h-
9	VII,21	Sp	VII,21	q+i-1	h-
10	D.1	h	Ac	$(q+i+M-2)_o$	
11	E.1	h-	Ac	$(q+i+h-2)_o$	
12	D.5	h	Ac	$(q+i+h-1)_o$	
13	VII,14	Sp	VII,14	q+i+h-1	
14	--				
[	q+i+h-1]		Ac	x_{i+h}	
15	--	h-			
[	q+i-1	h-]	Ac	$x_{i+h}-x_i$	
16	s.1	S	s.1	$x_{i+h}-x_i$	
17	F.2		Ac	$P_{i+1}^h(x)$	
18	F.1	h-	Ac	$P_{i+1}^h(x)-P_i^h(x)$	
19	s.1	+	R	$\frac{P_{i+1}^h(x)-P_i^h(x)}{x_{i+h}-x_i}$	
D.8	x				
VII,20	D.8		Ac		
21	--	h-			
[	q+i-1	h-]	Ac	$x-x_i$	
22	s.1	S	s.1	$x-x_i$	
23	s.1	x	Ac	$\frac{x-x_i}{x_{i+h}-x_i}(P_{i+1}^h(x)-P_i^h(x))$	
24	F.1	h	Ac	$P_i^{h+1}(x) = P_i^h(x) + \frac{x-x_i}{x_{i+h}-x_i}(P_{i+1}^h(x)-P_i^h(x))$	
25	--	S			
[	r+i-1	S]	C.i	$P_i^{h+i}(x)$	

(to VIII,1)

VIII,1	E.2		Ac	$(p+i)_0$	
2	D.5	h	Ac	$(p+i+1)_0$	
3	E.2	S	E.2	$(p+i+1)_0$	
4	E.1	h-	Ac	$(p+i+h-M+1)_0$	
5	D.5	h-	Ac	$(p+i+h-M)_0$	
6	D.4	h-	Ac	$(i+h-M)_0$	
7	X,1	Cc			
(to IX,1)					
IX,1	F.2		Ac	$p^h_{i+1}(x)$	
2	F.1	S	F.1	$p^h_{i+1}(x)$	
(to III,5)					
X,1	E.1		Ac	$(M-h)_0$	
2	D.5	h-	Ac	$(M-h-1)_0$	
3	E.1	S	E.1	$(M-h-1)_0$	
4	D.5	h-	Ac	$(M-h-2)_0$	
5	XI,1	Cc			
(to e)					
D.9	$(\alpha_2)_0$				
XI,1	D.9		Ac	$(\alpha_2)_0$	
2	II,4	Sp	II,4	α_2	C
D.10	$(\beta_2)_0$				
XI,3	D.10		Ac	$(\beta_2)_0$	
4	III,6	Sp	III,6	β_2	C
(to II,1)					

The ordering of the boxes is I, II; III; IV; V, VII, VIII, IX; X; XI; VI, and VII, e, II must also be the immediate successors of VI, X, XI, respectively, and III,4 and III,5 must be the immediate successors of IV and IX, respectively. This necessitates the extra orders

VI,6	VII,1	C
X,6	e	C
XI,5	II,1	C

and

IV,4	III,4	C
IX,3	III,5	C

α_1, α_2, β_1, β_2 correspond to III,1, IV,1, V,1, VI,1, respectively. Hence in the final enumeration III,1, IV,1 must have the same parity, and V,1, VI,1 must have the same parity.

We must now assign D.1-10, E.1-2, F.1-2, s.1 their actual values, pair the 75 orders I,1-6, II,1-4, III,1-6, IV,1-4, V,1-3, VI,1-6, VII,1-25, VIII,1-7, IX,1-3, X,1-6, XI,1-5 to 38 words, and then assign I,1-XI,5 their final values. These are expressed in this table:

I,1-6	0 -2'	VII,1-25	11'-23'	VI,1-6	35 -37'
II,1-4	3 -4'	VIII,1-7	24 -27	D.1-10	38 -47
III,1-6	5 -7'	IX,1-3	27'-28'	E.1-2	48 -49
IV,1-4	8 -9'	X,1-6	29 -31'	F.1-2	50 -51
V,1-3	10 -11	XI,1-5,*	32 -34'	s.1	52

Now we obtain this coded sequence:

0	38 ,	48 S	18	- ,	- h-	36	36 Sp',	-
1	39 ,	4 Sp'	19	52 S ,	51	37	51 S ,	11 C'
2	40 ,	7 Sp'	20	50 h-,	52 ÷	38	$(M-1)_o$	
3	41 ,	42 h	21	45 ,	- h-	39	5_o	
4	49 S ,	- C	22	52 S ,	52 x	40	10_o	
5	41 ,	6 Sp	23	50 h ,	- S	41	p_o	
6	- ,	50 S	24	49 ,	42 h	42	1_o	
7	49 ,	- C	25	49 S ,	48 h-	43	r_o	
8	43 ,	9 Sp	26	42 h-,	41 h-	44	$(q-p-1)_o$	
9	- ,	6 C'	27	29 Cc,	51	45	x	
10	10 Sp',	-	28	50 S ,	7 C	46	8_o	
11	51 S ,	49	29	48 ,	42 h-	47	35_o	
12	41 h-,	43 h	30	48 S ,	42 h-	48	-	
13	42 h-,	23 Sp'	31	32 Cc,	e C	49		
14	49 ,	44 h	32	46 ,	4 Sp'	50		
15	18 Sp',	21 Sp'	33	47 ,	7 Sp'	51		
16	38 h ,	48 h-	34	3 C ,	-	52		
17	42 h ,	18 Sp	35	41 h-,	43 h			

The durations may be estimated as follows:

I: 225 μ, II: 150 μ, III: 225 μ, IV: 275 μ, V: 125 μ, VI: 225 μ, VII: 1140 μ, VIII: 275 μ, IX: 200 μ, X: 225 μ, XI: 200 μ.

Total:
$$\mathrm{I} + \mathrm{II} \times (M-1) + \mathrm{III} + \mathrm{IV} \times (M-2) + \mathrm{V} \times (M-1) + \mathrm{VI} \times \frac{(M-1)(M-2)}{2} +$$
$$+ (\mathrm{VII} + \mathrm{VIII}) \frac{M(M-1)}{2} + \mathrm{IX} \times \frac{(M-1)(M-2)}{2} + \mathrm{X} \times (M-1) + \mathrm{XI} \times (M-2) =$$
$$= (225 + 150\,(M-1) + 225 + 275\,(M-2) + 125\,(M-1) + 225\,\frac{(M-1)(M-2)}{2} +$$
$$+ 1{,}415\,\frac{M(M-1)}{2} + 200\,\frac{(M-1)(M-2)}{2} + 225\,(M-1) + 200\,(M-2))\ \mu =$$
$$= (920\,M^2 - 370\,M - 575)\ \mu \approx (.9\,M^2 - .4\,M - .6)\ \mathrm{m}.$$

10.6 We now pass to the problem of interpolating a tabulated function based on the coding of Lagrange's interpolation formula in Problem 12.

PROBLEM 13.

The variable values $y_1, \ldots, y_N$ ($y_1 < y_2 < \ldots < y_N$) and the function values $q_1, \ldots, q_N$ are stored at two systems of N consecutive memory locations each: $\bar{q}, \bar{q}+1, \ldots, \bar{q}+N-1$ and $\bar{p}, \bar{p}+1, \ldots, \bar{p}+N-1$. The constants of the problem are $\bar{p}$, $\bar{q}$, N, M, y, to be stored at five given memory locations. It is desired to interpolate this function for the value y of the variable, using Lagrange's formula for the M points y_i nearest to y. ----

The problem should be treated differently, according to whether the $y_1, \ldots, y_N$ are or are not to be equidistant.

PROBLEM 13.a.

$y_1, \ldots, y_N$ are equidistant, i.e., $y_i = a + \frac{i-1}{N-1}(b-a)$. In this case only $y_1 = a$ and $y_N = b$ need be stored. ----

PROBLEM 13.b.

$y_1, \ldots, y_N$, are unrestricted. ----

In both cases our purpose is to reduce Problem 13 to Problem 12, with $x_1, \ldots, x_M$ equal to $y_k, \ldots, y_{k+M-1}$, and $p_1, \ldots, p_M$ equal to $q_k, \ldots, q_{k+M-1}$, where $k = 1, \ldots, N-M+1$ is so chosen that the $y_k, \ldots, y_{k+M-1}$ lie as close to y as possible.

10.7 We consider first Problem 13.a.

In this case the definition of k, as formulated at the end of 10.6, amounts to this: The remoter one of $y_k = a + \frac{k-1}{N-1}(b-a)$ and $y_{k+M-1} = a + \frac{k+M-2}{N-1}(b-a)$ should lie as close as possible to y. This is equivalent to requiring that their mean, $\frac{1}{2}(y_k + y_{k+M-1}) = a + \frac{2k+M-3}{2(N-1)}(b-a)$, lie as close as possible to y, i.e., that k lie as close as possible to $(N-1)\frac{y-a}{b-a} - \frac{M-3}{2}$.

There are various ways to find this k = k, based on iterative trial and error procedures. Since we will have to use a method of this type in connection with Problem 13.b, we prefer a different one at this occasion.

This method is based on the function $\{z\}$, which denotes the integer closest to z. Putting

$$k^{\bullet} = \left\{(N-1)\frac{y-a}{b-a} - \frac{M-3}{2}\right\},$$

we have clearly

$$\bar{k} \begin{cases} = 1 & \text{for } k^* \leq 1, \\ = k^* & \text{for } 1 \leq k \leq N\text{-}M\text{+}1, \\ = N\text{-}M\text{+}1 & \text{for } k^* \geq N\text{-}M\text{+}1 \end{cases}$$

Now the multiplication order (11, Table II), permits us to obtain $\{z\}$ directly. Indeed, the round off rule (cf. the discussion of the order in question) has the effect that when a product uv is formed, the accumulator will contain $\overrightarrow{uv} = 2^{-39}\{2^{39}uv\}$, and the arithmetical register will contain $\overrightarrow{\overrightarrow{uv}} = 2^{39}uv - \{2^{39}uv\}$. Hence putting $u = 2^{-39}(N-1)$ and $v = \frac{y-a}{b-a} - \frac{M-3}{2(N-1)}$ will produce $\overrightarrow{uv} = 2^{-39}\{2^{39}uv\} = 2^{-39}\bar{k}^*$ in the accumulator.

After k^* and $\bar{k}$ have been obtained, we can utilize the routine of Problem 12 to complete the task. This means, that we propose to use the coded sequence 0-52 of 10.5, and that we will adjust the coded sequence that will be formed here, so that it can be used in conjunction with that one of 10.5.

Among the constants of Problem 12 only p, q, and x need be given values which correspond to the new situation. x is clearly our present y. p,...,p+M-1 are the positions of the $p_1,\ldots,p_M$ of Problem 12, i.e., the positions of our present $q_{\bar{k}},\ldots,q_{\bar{k}+M-1}$, i.e., they are $\bar{p}+\bar{k}-1,\ldots,\bar{p}+\bar{k}+M-2$. Hence $p = \bar{p}+\bar{k}-1$. q,..., q+M-1 are the positions of the $x_1,\ldots,x_M$ of Problem 12, i.e., the positions of our present $y_{\bar{k}},\ldots,y_{\bar{k}+M-1}$. Since we determined in the formulation of Problem 13.a, that only $y_1 = a$ and $y_N = b$ will be stored, but not the entire sequence $y_1,\ldots,y_N$, this means that the desired sub-sequence $y_{\bar{k}},\ldots,y_{\bar{k}+M-1}$ does not exist anywhere ab initio.

Consequently q may have any value, all that is needed is that the positions q,...,q+M-1 should be available and empty (or irrelevantly occupied) when the coded sequence that we are going to formulate begins to operate. This sequence must then form

$$x_i = y_{\bar{k}+i-1} = a + \frac{\bar{k}+i-2}{N-1}(b-a)$$

and place it into the position q+i-1 for all i = 1,...,M.

It might seem wasteful to form x_i first, then store it at q+i-1, and finally obtain it from there by transfer when it is needed, i.e. during the period VII,1-19 of the coded sequence of 10.5. One might think that it is simpler to form x_i when it is actually needed, i.e., in VII,1-19 as stated above; the quantities needed are more specifically x_i, x_{i+h}, in the combination $\frac{x-x_i}{x_{i+h}-x_i}$), and thus avoid the transfers and the storage. It is easy to see, however, that the saving thus effected is altogether negligible, essentially because the size of our problem is proportional to M^2 (cf. the end of 10.5), while the number of steps required in forming and transferring the x_i's in the first mentioned way is only proportional to M. (Remember that M is likely to be ≥ 7, cf. the beginning of 10.5.) It does therefore hardly seem worthwhile to undertake those changes of the coded sequence of 10.5 which the second procedure would necessitate, and we will adhere to the first procedure.

In assigning letters to the various storage areas to be used, it must be remembered that the coded sequence that we are now developing is to be used in conjunction with (i.e., as a supplement to) the coded sequence of 10.5. The latter requires storage areas of various types: D-F, which are incorporated into its final enumeration (they are 38-51 of the 0-52 of 10.5); A (i.e., p,...,p+M-1), which will be part of our present A (i.e., of $\bar{p},\dots,\bar{p}+N-1$, it will be $\bar{p}+\bar{k}-1,\dots,\bar{p}+\bar{k}+M-2$; B (i.e., q,...,q+M-1), which may be at any available place; and C (i.e., r,...,r+M-2), which, too, may be at any available place. Therefore we can, in assigning letters to the various storage areas of the present coding, disregard those of 10.5, with the exception of B, C. Furthermore, there will be no need to refer here to C (of 10.5), since the coded sequence of 10.5 assumes the area C to be irrelevantly occupied. It will be necessary, however, to refer to B (of 10.5), since it is supposed to contain the $x_i = y_{\bar{k}+i-1}$ $(i = 1,\dots,M)$ of the problem. We will therefore think of the letters which are meant to designate storage areas of the coded sequence of 10.5 as being primed. This is, according to the above, an immediate necessity for B. We can now assign freely unprimed letters to the storage areas of the present coding.

Let A be the storage area corresponding to the interval from $\bar{p}$ to $\bar{p}+N-1$. In this way its positions will be A.1,...,N, where A.i corresponds to $\bar{p}+i-1$ and stores q_i. As in previous problems, the positions of A need not be shown in the final enumeration of the coded sequence.

The given data of the problem are $\bar{p}$, M, N, a, b, y, also the q, r of the coded sequence of 10.5. M, r are already stored there (as $(M-1)_0$, r_0 at 38, 43); and y, too (it coincides with x at 45). q, however, occurs only in combination with $\bar{k}$ (as $(q-p-1)_0 = (q-\bar{p}-\bar{k})_0$ at 44) and p, of course, contains $\bar{k}$ (as $p_0 = (\bar{p}+\bar{k}-1)_0$ at 41); and $\bar{k}$ originates in the machine. Hence, 41, 44 must be left empty (or, rather, irrelevantly occupied) when our present coded sequence begins to operate, and they must be appropriately substituted by its operations. q, however, must be stored as one of the constants of our present problem. Hence the constants requiring storage now (apart from M, r, y which we stored in 0-52, cf. above) are $\bar{p}$, q, N, a, b. They will be stored in the storage area B. (It will be convenient to store them as a, b, $(N-1)_0$, $(\bar{p}-1)_0$, $(q-1)_0$. We will also need 2^{-39} and 1_0; we will store the former in B and get the latter from 42.) The induction index i is a position mark; it will be stored as $(q+i-1)_0$ (i.e., $(B'.i)_0$) in the storage area C. The quantities which are processed during each inductive step will also be stored in the storage area C.

Our task is to calculate $\bar{k}$; to substitute $p_0 = (p+\bar{k}-1)_0$ and $(q-p-1)_0 = (\bar{q}-\bar{p}-k)_0$ into 41 and 44; and then to transfer $x_i = y_{\bar{k}+i-1}$ from A.$\bar{k}$+i-1 to B'.i. This latter operation is inductive. Finally, the control has to be sent, not to e, but to the beginning of 0-52, i.e., to 0.

We can now draw the flow diagram, as shown in Figure 10.4.

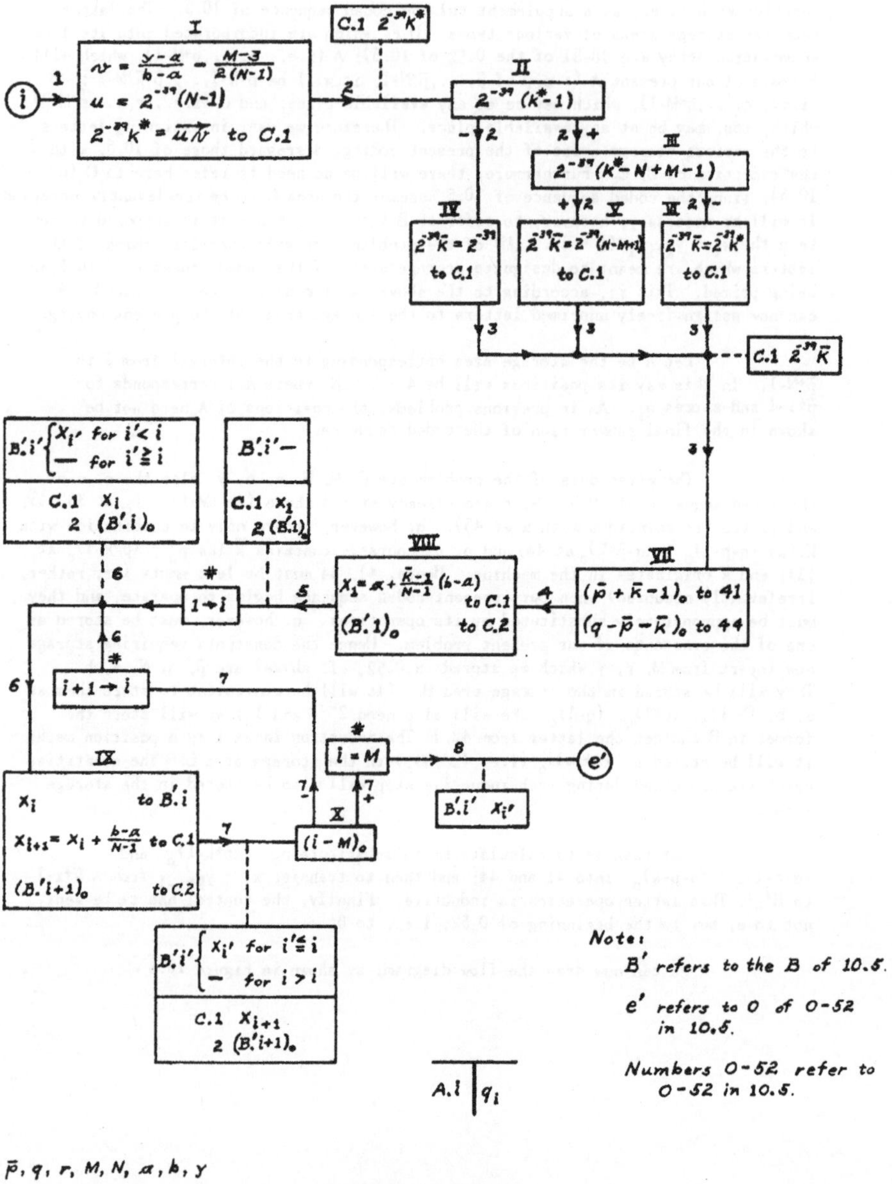
I
C.1 2^{-39}k*
ν = (y−α)/(b−α) − (M−3)/(2(N−1))
u = 2^{-39}(N−1)
2^{-39}k* = uν to C.1
II
2^{-39}(k*−1)
III
2^{-39}(k*−N+M−1)
IV
2^{-39}k̄ = 2^{-39}
to C.1
V
2^{-39}k̄ = 2^{-39}(N−M+1)
to C.1
VI
2^{-39}k̄ = 2^{-39}k*
to C.1
C.1 2^{-39}k̄
VII
(p̄ + k̄ − 1)_0 to 41
(q − p̄ − k̄)_0 to 44
VIII
x_1 = α + (k̄−1)/(N−1) (b−α) to C.1
(B′.1)_0 to C.2
B′.i′ { x_{i′} for i′ < i; — for i′ ≥ i
C.1 x_i
2 (B′.i)_0
B′.i′ —
C.1 x_1
2 (B.1)_0
1 → i
i+1 → i
i = M
e′
B′.i′ x_{i′}
IX
x_i to B′.i
x_{i+1} = x_i + (b−α)/(N−1) to C.1
(B′.i+1)_0 to C.2
X
(i − M)_0
B′.i′ { x_{i′} for i′ ≤ i; — for i′ > i
C.1 x_{i+1}
2 (B′.i+1)_0
A.1 q_i
p̄, q, r, M, N, α, b, y
i
Note:
B′ refers to the B of 10.5.
e′ refers to 0 of 0-52 in 10.5.
Numbers 0-52 refer to 0-52 in 10.5.

FIGURE 10.4

The actual coding will again deviate in some minor respects from the flow diagram, as in previous instances. In connection with this we will also need some extra storage capacity in C (C.1.1).

A matter of somewhat greater importance is this: We have noted before that since our machine recognizes numbers between -1 and 1 only, integers I must be stored in some other form. Frequently the form of a position mark, I_0, is per se more natural than any other (cf. 8.2); occasionally $2^{-39}I$ can be fitted more easily to the algebraical use to be made of I (cf. I in the present coding); sometimes $\frac{1}{I}$ is most convenient. (We could, of course, add to this list, but the three above forms seem to be the basic ones.) The transitions between these three forms are easily effected by using multiplications and divisions, but it seems natural to want to achieve the transitions between the two first ones (I_0 and $2^{-39}I$) in a more direct way.

This can be achieved by means of the partial substitution orders 18, 19 of Table II (x Sp, x Sp'), if they are modified as indicated in the Remark immediately preceding this chapter. We propose to use these orders now for arithmetical rather than logical (substitution) purposes, for positions x which contain no orders at all, but which are storing numbers in transit. Specifically: With I_0 in the accumulator, x Sp' produces $2^{-39}I$ at the position x; with $2^{-39}I$ in the accumulator x Sp produces $2^{-19}I$ at the position x, and a subsequent x h produces I_0 in the accumulator.

We mention that $\frac{1}{I}$ can be obtained from I_0 or $2^{-39}I$ by dividing them into 1_0 or 2^{-39}, respectively, and this without more than the usual loss in precision: Indeed, our division $\rho : \sigma$ is precise within an error 2^{-39}, no matter how small σ is (subject, of course, to the condition $|\sigma| > |\rho|$), provided that ρ, σ are given exactly. (If they are not given exactly, then their errors are amplified by $\frac{1}{\sigma}$, $\frac{\rho}{\sigma^2}$, respectively. In this case a small σ is dangerous, even though $|\sigma| > |\rho|$.) In the present case ρ, σ are given exactly. Forming 1_0, I_0, as well as 2^{-39}, $2^{-39}I$, involves no round-offs.

These methods will be used in our present coding: For transitions between I_0 and $2^{-39}I$, cf. I,22-23, III,3-4; VII,1-3; For forming reciprocals: $\frac{1}{N-1}$ from $(N-1)_0$ as $1_0 : (N-1)_0$ in VIII,4-5, and the very similar case of $\frac{M-3}{2(N-1)}$ from $(M-3)_0$, $(N-1)_0$ as $(M-3)_0 : 2(N-1)_0$ in I,9-15. Regarding $\frac{1}{N-1}$ we also note this: We need $\frac{b-a}{N-1}$ in VIII. We will form $\frac{1}{N-1} = 1_0 : (N-1)_0$ first and $\frac{b-a}{N-1} = \frac{1}{N-1} \times (b-a)$ afterwards. Forming $1_0 \times (b-a)$ first and $\frac{b-a}{N-1} = [1_0 \times (b-a)] : (N-1)_0$ afterwards would lead to a serious loss of precision, since $1_0 \times (b-a)$ plays the role of ρ above, and as it involves a round-off it is not given exactly, and hence may cause a loss of precision as indicated there.

We will have to refer in our present coding repeatedly to 0-52 in 10.5. It should therefore be remembered that this coded sequence is now supposed to be changed insofar that 41, 44 are irrelevant and 45 contains y. 38, 43 contain $(M-1)_0$, r_0, as in 10.5.

The static coding of the boxes I-X follows:

B.1	a			
2	b			
I,1	B.2		Ac	b
2	B.1	h-	Ac	b-a
3	s.1	S	s.1	b-a
4	45		Ac	y
5	B.1	h-	Ac	y-a
6	s.1	÷	R	$\frac{y-a}{b-a}$
7		n	Ac	$\frac{y-a}{b-a}$
8	s.1	S	s.1	$\frac{y-a}{b-a}$
B.3	$(N-1)_o$			
I,9	B.3		Ac	$(N-1)_o$
10	B.3	h	Ac	$2(N-1)_o$
11	s.2	S	s.2	$2(N-1)_o$
12	38		Ac	$(M-1)_o$
13	42	h-	Ac	$(M-2)_o$
14	42	h-	Ac	$(M-3)_o$
15	s.2		R	$\frac{M-3}{2(N-1)} = \frac{(M-3)_o}{2(N-1)_o}$
16			Ac	$\frac{M-3}{2(N-1)}$
17	s.2	S	s.2	$\frac{M-3}{2(N-1)}$
18	s.1		Ac	$\frac{y-a}{b-a}$
19	s.2	h-	Ac	$v = \frac{y-a}{b-a} - \frac{M-3}{2(N-1)}$
20	s.1	S	s.1	
21	s.1	R	R	v
22	B.3		Ac	$(N-1)_o$
23	s.2	Sp'	s.2	$u = 2^{-39}(N-1)$
24	s.2	x	Ac	$2^{-39}k^* = \overrightarrow{uv}$
25	C.1	S	C.1	$2^{-39}k^*$
(to II,1)				
B.4	2^{-39}			
II,1	C.1		Ac	$2^{-39}k^*$
2	B.4	h-	Ac	$2^{-39}(k^*-1)$
3	III,1	Cc		
(to IV,1)				

III,1	B.3		Ac	$(N-1)_o$
2	38	h-	Ac	$(N-M)_o$
3	42	h	Ac	$(N-M+1)_o$
4	C.1.1	Sp'	C.1.1	$2^{-39}(N-M+1)$
5	C.1		Ac	$2^{-39}k^*$
6	C.1.1	h-	Ac	$2^{-39}(k^*-N+M-1)$
7	V,1	Cc		
(to VI,1)				
IV,1	B.4		Ac	$2^{-39}\bar{k} = 2^{-39}$
(to V,2)				
V,1	C.1.1		Ac	$2^{-39}\bar{k} = 2^{-39}(N-M+1)$
2	C.1	S	C.1	$2^{-39}\bar{k}$
(to VII,1)				
VI,				
(to VII,1)				
VII,1	C.1		Ac	$2^{-39}\bar{k}$
2	s.1	Sp	s.1	$2^{-19}\bar{k}$
3	s.1	h	Ac	$\bar{k}_o = 2^{-19}\bar{k}+2^{-39}\bar{k}$
B.5	$(\bar{p}-1)_o$			
VII,4	B.5	h	Ac	$(\bar{p}+\bar{k}-1)_o$
5	41	S	41	$(\bar{p}+\bar{k}-1)_o$
6	s.1	S	s.1	$(\bar{p}+\bar{k}-1)_o$
B.6	$(q-1)_o$			
VII,7	B.6		Ac	$(q-1)_o$
8	s.1	h-	Ac	$(q-\bar{p}-\bar{k})_o$
9	44	S	44	$(q-\bar{p}-\bar{k})_c$
(to VIII,1)				
VIII,1	B.2		Ac	b
2	B.1	h-	Ac	b-a
3	s.1	S	s.1	b-a
	42		Ac	1_o
5	B.3		R	$\frac{1}{N-1} = \frac{1_o}{(N-1)_o}$
6	s.1		Ac	$\frac{b-a}{N-1}$
	C.1.1	S	C.1.1	$\frac{b-a}{N-1}$
8	C.1.1	R	R	$\frac{b-a}{N-1}$
9	C.1		Ac	$2^{-39}\bar{k}$
10	B.4	h-	Ac	$2^{-39}(\bar{k}-1)$

VIII,11	s.1	S	s.1	$2^{-39}(\bar{k}-1)$	
12	s.1	⌄	R	$\frac{\bar{k}-1}{N-1}(b-a)$	
13		A	Ac	$\frac{\bar{k}-1}{N-1}(b-a)$	
14	B.1	h	Ac	$x_1 = a + \frac{\bar{k}-1}{N-1}(b-a)$	
15	C.1	S	C.1	x_1	
16	B.6		Ac	$(q-1)_0$	
17	42	h	Ac	q_0	
18	C.2	S	C.2	q_0	
(to IX,1)					
IX,1	C.2		Ac	$(q+i-1)_0$	
2	IX,6	Sp	IX,6	$q+i-1$	S
3	42	h	Ac	$(q+i)_0$	
4	C.2	S	C.2	$(q+i)_0$	
5	C.1		Ac	x_i	
6	-	S			
[	q+i-1	S]	B'.i	x_i	
	C.1.1	h	Ac	$x_{i+1} = x_i + \frac{b-a}{N-1}$	
8	C.1	S	C.1	x_{i+1}	
(to X,1)					
X,1	C.2		Ac	$(q+i)_0$	
2	38	h-	Ac	$(q+i-M+1)_0$	
3	B.6	h-	Ac	$(i-M+2)_0$	
4	42	h-	Ac	$(i-M+1)_0$	
5	42	h-	Ac	$(i-M)_0$	
6	e'	Cc			
(to IX,1)					

Note, that the box VI required no coding, hence its immediate successor (VII) must follow directly upon its immediate predecessor (III).

The ordering of the boxes is I, II, IV; III, VII, VIII, IX, X; V and VII, IX must also be the immediate successors of V, X, respectively, and V,2 must be the immediate successor of IV. This necessitates the extra orders

V,3	VII,1	C
X,7	IX,1	C

and

IV,2	V,2	C

As indicated in Figure 10.4, e' is 0.

We must now assign B.1-6, C.1-2, 1.1, s.1-2 their actual values, pair the 82 orders I,1-25, II,1-3, III,1-7, IV,1-2, V,1-3, VII,1-9, VIII,1-18, IX,1-8, X,1-7 to 41 words, and then assign I,1-X,7 their actual values. We wish to do this as a continuation of the code of 10.5. We will therefore begin with the number 53. Furthermore the contents of C.1-2, C.1.1, s.1-2 are irrelevant like those of 48-52 there. Hence they may be made to coincide with these. We therefore identify them accordingly . Summing all these things up, we obtain the following table:

I,1-25	53 -65	VII,1-9	71'-75'	V,1-3	92'-93'
II,1-3	65'-66'	VIII,1-18	76 -84'	B.1-6	94 -99
IV,1-2	67 -67'	IX,1-8	85 -88'	C.1-2	48 -49
III,1-7	68 -71	X,1-7	89 -92	C.1.1	50
				s.1-2	51-52

Now we obtain this coded sequence:

53	95 ,	94 h-	69	42 h ,	50 Sp'	85	49 ,	87 Sp'
54	51 S ,	45	70	48 ,	50 h-	86	42 h ,	49 S
55	94 h-,	51 ÷	71	92 Cc',	48	87	48 ,	- S
56	A ,	51 S	72	51 Sp ,	51 h	88	50 h ,	48 S
57	96 ,	96 h	73	98 h ,	41 S	89	49 ,	38 h-
58	52 S ,	38	74	51 S ,	99	90	99 h-,	42 h-
59	42 h-,	42 h-	75	51 h- ,	44 S	91	42 h-,	0 Cc
60	52 ÷ ,	A	76	95 ,	94 h-	92	85 C ,	50
61	52 S ,	51	77	51 S ,	42	93	48 S ,	71 C'
62	52 h-,	51 S	78	96 ÷ ,	51 x	94	a	
63	51 R ,	96	79	50 S ,	50 R	95	b	
64	52 Sp',	52 x	80	48 ,	97 h-	96	$(N-1)_0$	
65	48 S ,	48	81	51 S ,	51 x	97	2^{-39}	
66	97 h-,	68 Cc	82	A ,	94 h	98	$(\bar{p}-1)_0$	
67	97 ,	93 C	83	48 S ,	99	99	$(q-1)_0$	
68	96 ,	38 h-	84	42 h ,	49 S			

For the sake of completeness, we restate that part of 0-52 of 10.5 which contains all changes and all substitutable constants of the problem. This is 38-52:

38	$(M-1)_0$	43	r_0	48	--
39	5_0	44	--	49	--
40	10_0	45	y	50	--
41	--	46	8_0	51	--
42	1_0	47	35_0	52	--

The durations may be estimated as follows:

I: 1,220 μ, II: 125 μ, III: 275 μ, IV: 150 μ, V: 125 μ, VII: 350 μ, VIII: 915 μ, IX: 300 μ, X: 275 μ.

Total: I + II + (IV or (III + V) or III) + VII + VIII + (IX + X) × M =

maximum = (1,220 + 125 + 400 + 350 + 915 + 575 M) μ =

= (575 M + 3,010) $\mu \approx$ (.6 M + 3) m.

Hence the complete interpolation procedure (10.5 plus 10.7) requires

$$(.9\, M^2 + .2\, M + 2.4)\ m.$$

10.8 We consider next Problem 13.b. We are again looking for that $k = 1,\ldots,N-M+1$, for which $y_k,\ldots,y_{k+M-1}$ lie as close to y as possible (cf. the end of 10.6), i.e., for which the remoter one of y_k and y_{k+M-1} lies as close to y as possible.

Since $y_1,\ldots,y_N$ are not equidistant, we cannot find k by an arithmetical criterium, as in 10.7. We must proceed by trial and error, and we will now describe a method which achieves this particularly efficiently.

We are then looking for a $k = \bar{k}$ which minimizes $\mu_k = \text{Max}\,(y-y_k,\ y_{k+M-1}-y)$ in $k = 1,\ldots,N-M+1$.

Let us first note this: $y-y_k > y-y_{k+1}$, hence $y-y_k > y_{k+M}-y$ implies $y-y_k > \mu_{k+1}$, and therefore $\mu_k > \mu_{k+1}$. On the other hand $y_{k+M-1}-y < y_{k+M}-y$, hence $y-y_k \leqq y_{k+M}-y$ implies $\mu_k \leqq y_{k+M}-y$, and therefore $\mu_k \leqq \mu_{k+1}$. Hence $\mu_k >$ or $\leqq \mu_{k+1}$, according to whether $y-y_k >$ or $\leqq y_{k+M}-y$ i.e., $y_k + y_{k+M} - 2y <$ or $\geqq 0$.

In order to keep the size between -1 and 1, it is best to replace $y_k + y_{k+M} - 2y$ by

(1) $$z_k = \frac{1}{2}\, y_k + \frac{1}{2}\, y_{k+M} - y .$$

z_k is monotone increasing in k. Therefore $z_k < 0$ implies $z_h < 0$ and hence $\mu_h > \mu_{h+1}$ for all $h \leqq k$, i.e., $\mu_1 > \mu_2 > \ldots > \mu_k > \mu_{k+1}$. Similarly $z_k \geqq 0$ implies $z_h \geqq 0$ and hence $\mu_h \leqq \mu_{h+1}$ for all $h \geqq k$, i.e., $\mu_k \leqq \mu_{k+1} \leqq \ldots \leqq \mu_{N-M} \leqq \mu_{N-M+1}$. From these we may infer

(2)
$$z_k < 0 \ \text{ implies } \ \bar{k} > k,$$
$$z_k \geqq 0 \ \text{ implies that we can choose } \bar{k} \leqq k.$$

Consequently we can obtain $\bar{k}$ by "bracketing" guided by the sign of z_k.

Note that z_k can be formed for $k = 1,\ldots,N-M$ only, but not for $k = N-M+1$. The "bracketing" must begin by testing the sign of z_k for $k = 1$, if it is +, then $\bar{k} = 1$, if it is -, then we continue. Next we test the sign of z_k for $k = N-M$, if it is -, then $\bar{k} = N-M+1$, if it is +, then we continue. In this case we know that $1 < \bar{k} \leqq N-M$. Put $k_1^- = 1$, $k_1^+ = N-M$. Consider more generally the case where we know that $k_i^- < \bar{k} \leqq k_i^+$. This implies $k_i^+-k_i^- \geqq 1$. If $k_1^+-k_1^- = 1$, then $\bar{k} = k_1^+$, if $k_1^+-k_1^- > 1$, then we continue.

In this case we use the function $[w]$ which denotes the largest integer $\leq w$. Put

$$k_i^o = \left[\frac{1}{2} (k_i^+ + k_i^-) \right] , \tag{3}$$

and test the sign of z_k for $k = k_i^o$, if it is + then $\bar{k} \leq k_i^o$, if it is - then $\bar{k} > k_i^o$. We can therefore put $k_{i+1}^- = k_i^-$, $k_{i+1}^+ = k_i^o$ or $k_{i+1}^- = k_i^o$, $k_{i+1}^+ = k_i^+$, respectively. This completes the inductive step from i to i+1. Sooner or later, say for $i = i_o$, $k_i^+ - k_i^- = 1$ will occur, and the process will terminate.

Note, that $k_{i+1}^+ - k_{i+1}^- \leq \frac{1}{2}(k_i^+ - k_i^- + 1)$, i.e., $k_i^+ - k_i^- \geq 2(k_{i+1}^+ - k_{i+1}^-) - 1$, i.e., $k_{i-1}^+ - k_{i-1}^- \geq 2(k_i^+ - k_i^-) - 1$. For $i = i_o$, however $k_i^+ - k_i^- = 1$, $k_{i-1}^+ - k_{i-1}^- \neq 1$, hence $k_{i-1}^+ - k_{i-1}^- \geq 2$. Thus $k_{i_o-1}^+ - k_{i_o-1}^- \geq 2$, $k_{i_o-2}^+ - k_{i_o-2}^- \geq 3, \ldots, k_1^+ - k_1^- \geq 2^{i_o-2}+1$. However, $k_1^+ - k_1^- = N-M-1$, therefore,

$$i_o \leq {}^2\log (N-M-2) + 2. \tag{4}$$

The virtue of this "bracketing" method is, of course, that the number of steps it requires is of the order of ${}^2\log N$, and not, as it would be with most other trial and error methods, of the order of N. (Note, that N is likely to be large compared to M, which alone figures in the estimates of 10.5 and 10.7.)

After $\bar{k}$ has been obtained, we can utilize the routine of Problem 12 to complete the task, just as at the corresponding point of the discussion of Problem 13.a in 10.7. This means that we propose to use the coded sequence 0-52 of 10.5, and that we will adjust the coded sequence which will be formed here, just as in 10.7.

The situation with the constants of Problem 12 is somewhat simpler than it was in 10.7. Again, p, q, and x need be given values which correspond to the new situation. x is clearly our present y. $p, \ldots, p+M-1$ and $q, \ldots, q+M-1$ are the positions of the $p_1, \ldots, p_M$ and $x_1, \ldots, x_M$ of Problem 12, i.e., the positions of our present $q_{\bar{k}}, \ldots, q_{\bar{k}+M-1}$ and $y_{\bar{k}}, \ldots, y_{\bar{k}+M-1}$, i.e., they are $\bar{p}+\bar{k}-1, \ldots, \bar{p}+\bar{k}+M-2$ and $\bar{q}+\bar{k}-1, \ldots, \bar{q}+\bar{k}+M-2$. Hence $p = \bar{p}+\bar{k}-1$, $q = \bar{q}+\bar{k}-1$. The complications in 10.7, connected with q, or, more precisely, with the $y_{\bar{k}}, \ldots, y_{\bar{k}+M-1}$, do not arise here, since Problem 13.b provided for storing the entire sequence $y_1, \ldots, y_N$.

In assigning letters to the various storage areas to be used, it must be remembered, just as at the corresponding point in 10.7, that the coded sequence that we are now developing is to be used in conjunction with (i.e., as a supplement to) the coded sequence of 10.5. As in 10.7, we have to classify the storage areas required by the latter, but this classification now differs somewhat from that one of 10.7: We have the storage areas D-F, which are incorporated in the final enumeration (they are 38-51 of the 0-52 of 10.5); A and B (i.e., $p, \ldots, p+M-1$ and $q, \ldots, q+M-1$), which will be part of our present A and B (i.e., of $\bar{p}, \ldots, \bar{p}+N-1$ and $\bar{q}, \ldots, \bar{q}+N-1$, they will be $\bar{p}+\bar{k}-1, \ldots, \bar{p}+\bar{k}+M-2$ and $\bar{q}+\bar{k}-1, \ldots, \bar{q}+\bar{k}+M-2$); and C (i.e., $r, \ldots, r+M-2$), which may be at any available place. Therefore, we can, in assigning letters to the various storage areas in the present coding, disregard those of 10.5, with the exception of C. Furthermore, there will again be no need to refer here to C (of 10.5), since the coded sequence of 10.5 assumes the area C to be irrelevantly occupied. We will, at any rate, think of the letters which are meant to designate storage areas of the coded sequence of 10.5 as being primed. We can now assign freely unprimed letters to the storage areas of the present coding.

Let A and B be the storage areas corresponding to the intervals from $\bar{p}$ to $\bar{p}$+N-1 and from $\bar{q}$ to $\bar{q}$+N-1. In this way their positions will be A.1,...,N and B.1,...,N, where A.i and B.i correspond to $\bar{p}$+i-1 and $\bar{q}$+i-1 and store q_i and y_i. As in previous problems, the position of A and B need not be shown in the final enumeration of the coded sequence.

The given data of the problem are $\bar{p}$, $\bar{q}$, M, N, y, also r of the coded sequence of 10.5. M, r are already stored there (as $(M-1)_0$, r_0 at 38, 43); and y, too (it coincides with x at 45). The definitions of p, q involve $\bar{k}$ ($p=\bar{p}+\bar{k}-1$, $q=\bar{q}+\bar{k}-1$), and $\bar{k}$ originates in the machine. The form in which p is stored accordingly involves $\bar{k}$ (as $p_0 = (\bar{p}+\bar{k}-1)_0$ at 41); while the form in which q is stored does not happen to involve $\bar{k}$ (as $(q-p-1)_0 = (\bar{q}-\bar{p}-1)_0$ at 44). Hence 41 must be left empty (or rather, irrelevent1y occupied) when our present coded sequence begins to operate, and it must be appropriately substituted by its operation. 44 might be used to store $(\bar{q}-\bar{p}-1)_0$ from the start, but we prefer to leave it, too, empty (i.e., irrelevantly occupied) and to substitute it in the process. Hence the constants requiring storage now (apart from M, r, y which are stored in 0-52, cf. above) are $\bar{p}$, $\bar{q}$, N. They will be stored in the storage area C. (It will be convenient to store them as $(\bar{p}-1)_0$, $(\bar{q}-1)_0$, $(N-1)_0$.) The exit-locations of the variable remote connections will also be accommodated in C. The induction index i need not be stored explicitly, since the quantities k_i^o, k_i^-, k_i^+ contain all that is needed to keep track of the progress of the induction. (The $i = i_0$ for which the "bracketing", i.e., the induction, ends, is defined by $k_i^+ - k_i^- = 1$, cf. above.) k_i^o, k_i^-, k_i^+ are position marks, they will be stored as $(\bar{q}+k_i^o-1)_0$, $(\bar{q}+k_i^--1)_0$, $(q+k_i^+-1)_0$ (i.e., $(B.k_i^o)_0$, $(B.k_i^-)_0$ $(B.k_i^+)_0$) in the storage area D. Finally, k, when formed (in the form $(\bar{q}+\bar{k}-1)_0$, i.e., $(B.\bar{k})_0$) , will be stored in D, replacing the corresponding expression of k_i^+.

The index i runs from 1 to i_0. In drawing the flow diagram, it is convenient to treat the two preliminary steps (the sensing of the sign of z_k for k = 1 and k = N-M) as part of the same induction, and associate them ideally with two values of i preceding i = 1, i.e., with i = -1 and i = 0. For i = -1, 0 k_i^o may be defined as the value of k for which the sign of z_k is being tested (this is its role for $i \geqq 1$), i.e., as 1, N-M, respectively; while k_i^-, k_i^+ may remain undefined for i = -1, 0.

Our task is to calculate k (in the form $(\bar{q}+\bar{k}-1)_0$, i.e., $(B.\bar{k})_c$); to substitute $p_0 = (\bar{p}+\bar{k}-1)_0$ and $(q-p-1)_0 = (\bar{q}-\bar{p}-1)_0$ into 41 and 44; and finally to send the control, not to e, but to the beginning of 0-52, i.e. to 0.

We can now draw the flow diagram, as shown in Figure 10.5. The variable remote connections α and β are necessary in order to make the differing treatments for i = -1, for i = 0, and for $i \geqq 1$, possible (cf. above).

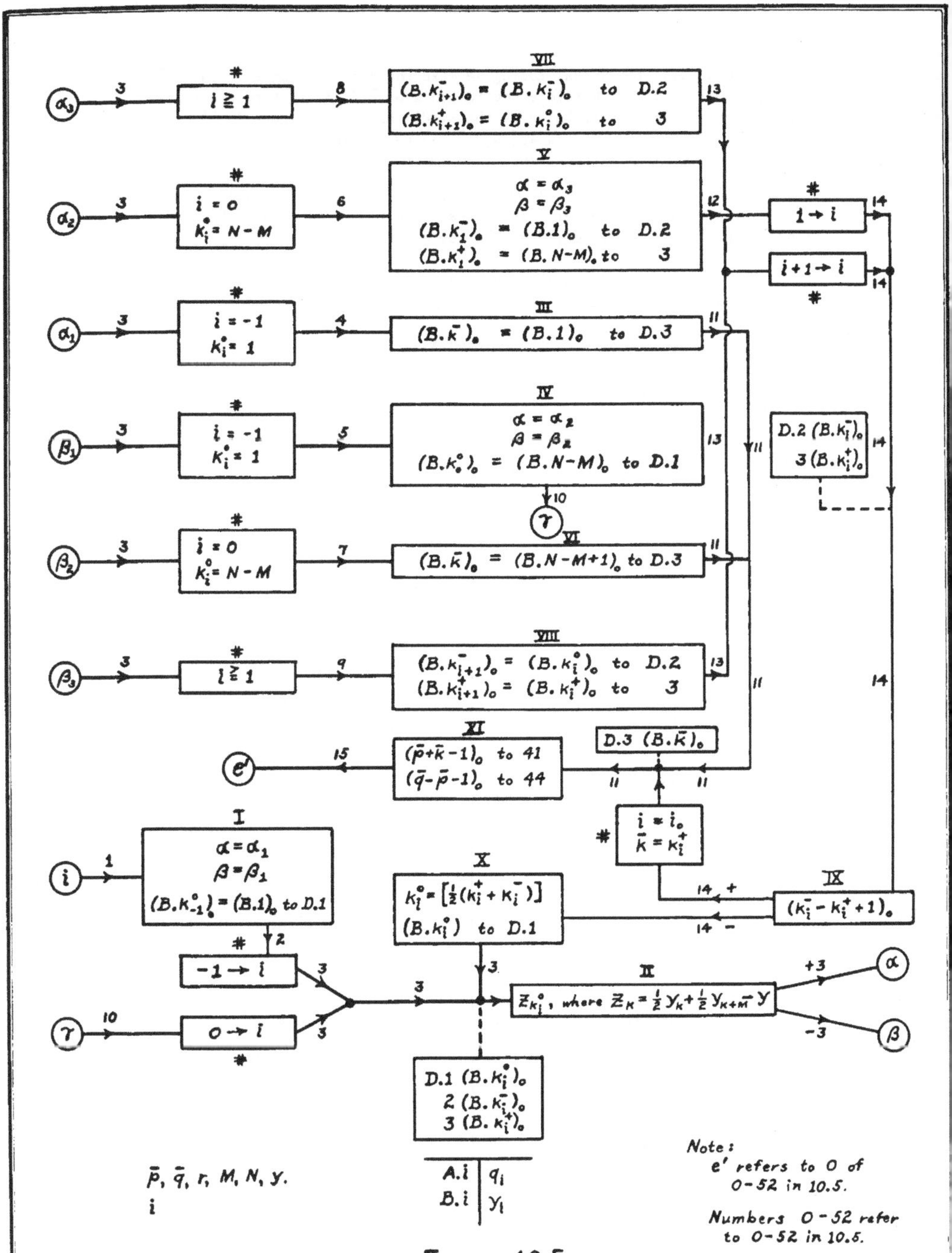

VII
$(B.k^-_{i+1})_0 = (B.k^-_i)_0$ to D.2
$(B.k^+_{i+1})_0 = (B.k^0_i)_0$ to 3
V
$\alpha = \alpha_3$
$\beta = \beta_3$
$(B.k^-_1)_0 = (B.1)_0$ to D.2
$(B.k^+_i)_0 = (B.N-M)_0$ to 3
III
$(B.\bar{k})_0 = (B.1)_0$ to D.3
IV
$\alpha = \alpha_2$
$\beta = \beta_2$
$(B.k^0_0)_0 = (B.N-M)_0$ to D.1
VI
$(B.\bar{k})_0 = (B.N-M+1)_0$ to D.3
VIII
$(B.k^-_{i+1})_0 = (B.k^0_i)_0$ to D.2
$(B.k^+_{i+1})_0 = (B.k^+_i)_0$ to 3
$i \geq 1$
$i = 0$, $k^0_i = N-M$
$i = -1$, $k^0_i = 1$
$1 \to i$
$i+1 \to i$
D.2 $(B.k^-_i)_0$
3 $(B.k^+_i)_0$
D.3 $(B.\bar{k})_0$
XI
$(\bar{p}+\bar{k}-1)_0$ to 41
$(\bar{q}-\bar{p}-1)_0$ to 44
I
$\alpha = \alpha_1$
$\beta = \beta_1$
$(B.k^0_{-1})_0 = (B.1)_0$ to D.1
$-1 \to i$
$0 \to i$
$i = i_0$
$\bar{k} = k^+_i$
X
$k^0_i = [\frac{1}{2}(k^+_i + k^-_i)]$
$(B.k^0_i)$ to D.1
IX
$(k^-_i - k^+_i + 1)_0$
II
$Z_{k^0_i}$, where $Z_K = \frac{1}{2}y_K + \frac{1}{2}y_{K+M} - y$
D.1 $(B.k^0_i)_0$
2 $(B.k^-_i)_0$
3 $(B.k^+_i)_0$
$\bar{p}, \bar{q}, r, M, N, y.$
i
A.i | q_i
B.i | y_i
Note:
e' refers to 0 of 0-52 in 10.5.
Numbers 0-52 refer to 0-52 in 10.5.

FIGURE 10.5

Regarding the actual coding we make these observations: Since k_i^-, k_i^+ are undefined for i = -1, 0, therefore the contents of D.2, 3 are irrelevant for i = -1, 0 (cf. the storage distributed at the middle bottom of Figure 10.5). The forming of $(B.k_i^o)$ in X, with $k_i^o = [\frac{1}{2}(k_i^+ + k_i^-)]$, is best effected by detouring over the quantities $2^{-39}(\bar{q}+k_i^o-1)$, $2^{-39}(\bar{q}+k_i^- -1)$, $2^{-39}(\bar{q}+k_i^+ -1)$, because the operation $m = [\frac{1}{2} n]$ is most easily performed with the help of $2^{-39}m$, $2^{-39}n$: Indeed the right-shift R carries the latter directly into the former. The transitions between I_o and $2^{-39}I$ (in both directions) are performed as discussed at the corresponding place in 10.7.

We will use 0-52 in 10.5 just as in 10.7, and the remarks which we made there on this subject apply again.

The static coding of the boxes I-XI follows:

C.1	$(\alpha_1)_o$				
2	$(\beta_1)_o$				
I,1	C.1		Ac	$(\alpha_1)_o$	
2	II,13	Sp	II,13	α_1	Cc
3	C.2		Ac	$(\beta_1)_o$	
4	II,14	Sp	II,14	β_1	C
C.3	$(\bar{q}-1)_o$				
I,5	C.3		Ac	$(\bar{q}-1)_o$	
6	42	h	Ac	$\bar{q}_o$	
7	D.1	S	D.1	$\bar{q}_o$	
(to II,1)					
II,1	D.1		Ac	$(\bar{q}+k_i^o-1)_o$	
2	II,6	Sp	II,6	$\bar{q}+k_i^o-1$	
3	38	h	Ac	$(\bar{q}+k_i^o+M-2)_o$	
4	42	h	Ac	$(\bar{q}+k_i^o+M-1)_o$	
5	II,9	Sp	II,9	$\bar{q}+k_i^o+M-1$	
6	--				
[	$\bar{q}+k_i^o-1$]		Ac	$y_{k_i^o}$	
		R	Ac	$\frac{1}{2} y_{k_i^o}$	
8	s.1	S	s.1	$\frac{1}{2} y_{k_i^o}$	
9	--				
[	$\bar{q}+k_i^o+M-1$]		Ac	$y_{k_i^o+M}$	
10		R	Ac	$\frac{1}{2} y_{k_i^o+M}$	
11	s.1	h	Ac	$\frac{1}{2} y_{k_i^o} + \frac{1}{2} y_{k_i^o+M}$	
12	45	h-	Ac	$z_{k_i^o} = \frac{1}{2} y_{k_i^o} + \frac{1}{2} y_{k_i^o+M} - y$	

II,13	--	Cc				
[	α	Cc]				
14	--	C				
[	β	C]				
III,1	C.3			Ac	$(\bar{q}-1)_0$	
2	42	h		Ac	$\bar{q}_0$	
3	D.3	S		D.3	$\bar{q}_0$	
(to XI,1)						
C.4	$(\alpha_2)_0$					
5	$(\beta_2)_0$					
IV,1	C.4			Ac	$(\alpha_2)_0$	
2	II,13	Sp		II,13	α_2	Cc
3	C.5			Ac	$(\beta_2)_0$	
4	II,14	Sp		II,14	β_2	C
C.6	$(N-1)_0$					
IV,5	C.6			Ac	$(N-1)_0$	
6	38	h-		Ac	$(N-M)_0$	
7	C.3	h		Ac	$(\bar{q}+N-M-1)_0$	
8	D.1	S		D.1	$(\bar{q}+N-M-1)_0$	
(to II,1)						
C.7	$(\alpha_3)_0$					
8	$(\beta_3)_0$					
V,1	C.7			Ac	$(\alpha_3)_0$	
2	II,13	Sp		II,13	α_3	Cc
3	C.8			Ac	$(\beta_3)_0$	
4	II,14	Sp		II,14	β_3	C
5	C.3			Ac	$(\bar{q}-1)_0$	
6	42	h		Ac	$\bar{q}_0$	
7	D.2	S		D.2	$\bar{q}_0$	
8	C.3			Ac	$(\bar{q}-1)_0$	
9	C.6	h		Ac	$(\bar{q}+N-2)_0$	
10	38	h-		Ac	$(\bar{q}+N-M-1)_0$	
11	D.3	S		D.3	$(\bar{q}+N-M-1)_0$	
(to IX,1)						
VI,1	C.3			Ac	$(\bar{q}-1)_0$	
2	C.6	h		Ac	$(\bar{q}+N-2)_0$	
3	38	h-		Ac	$(\bar{q}+N-M-1)_0$	
4	42	h		Ac	$(\bar{q}+N-M)_0$	
5	D.3	S		D.3	$(\bar{q}+N-M)_0$	
(to XI,1)						

VII,1	D.1		Ac	$(\bar{q}+k_i^0-1)_0$
2	D.3	S	D.3	$(\bar{q}+k_i^0-1)_0$
(to IX,1)				
VIII,1	D.1		Ac	$(\bar{q}+k_i^0-1)_0$
2	D.2	S	D.2	$(\bar{q}+k_i^0-1)_0$
(to IX,1)				
IX,1	D.2		Ac	$(\bar{q}+k_i^--1)_0$
2	D.3	h-	Ac	$(k_i^--k_i^+)_0$
3	42	h	Ac	$(k_i^--k_i^++1)_0$
4	XI,1	Cc		
(to X,1)				
X,1	D.2		Ac	$(\bar{q}+k_i^--1)_0$
2	D.3	h	Ac	$(2\bar{q}+k_i^++k_i^--2)_0$
3	s.1	Sp'	s.1	$2^{-39}(2\bar{q}+k_i^++k_i^--2)$
4	s.1		Ac	$2^{-39}(2\bar{q}+k_i^++k_i^--2)$
5		R	Ac	$2^{-39}(q+k_i^0-1) =$
				$= 2^{-39}(\bar{q}+\frac{1}{2}[k_i^++k_i^-]-1)$
6	s.1	Sp	s.1	$2^{-19}(\bar{q}+k_i^0-1)$
7	s.1	h	Ac	$(\bar{q}+k_i^0-1)_0$
8	D.1	S	D.1	$(\bar{q}+k_i^0-1)_0$
(to II,1)				
C.9	$(\bar{p}-1)_0$			
XI,1	C.9		Ac	$(\bar{p}-1)_0$
2	C.3	h-	Ac	$(\bar{p}-\bar{q})_0$
3	D.3	h	Ac	$(\bar{p}+\bar{k}-1)_0$
4	41	S	41	$(\bar{p}+\bar{k}-1)_0$
5	C.3		Ac	$(\bar{q}-1)_0$
6	C.9	h-	Ac	$(\bar{q}-\bar{p})_0$
7	42	h-	Ac	$(\bar{q}-\bar{p}-1)_0$
8	44	S	44	$(\bar{q}-\bar{p}-1)_0$
9	e'	C		

The ordering of the boxes is I, II; III, XI; IV; V, IX, X; VI; VII; VIII, and II, XI, IX, IX, II must also be the immediate successors of IV, VI, VII, VIII, X, respectively. This necessitates the extra orders

IV,9	II,1	C
VI,6	XI,1	C
VII,3	IX,1	C
VIII,3	IX,1	C
X,9	II,1	C

As indicated in Figure 10.5, e' is 0.

α_1, α_2, α_3, β_1, β_2, β_3 correspond to III,1, V,1, VII,1, IV,1, VI,1, VIII,1, respectively. Hence in the final enumeration III,1, V,1, VII,1 must have the same parity and IV,1, VI,1, VIII,1, must have the same parity.

We must now assign C.1-9, D.1-3, s.1 their actual values, pair the 78 orders I,1-7, II,1-14, III,1-3, IV,1-9, V,I-11, VI,I-6, VII,1-3, VIII,1-3, IX,1-4, X,1-9, XI,1-9 to 39 words (actually two dummy orders, necessitated by the adjusting of the parities of V,1 and VIII,1 to those of III,1 and IV,1, respectively, increase this to 40) and then assign I,I-XI,9 their actual values. We wish to do this again as a continuation of the code of 10.5. We will therefore again begin with the number 53. Furthermore the contents of D.1-3, s.1 are irrelevant like those of 48-52 there. Hence they may be made to coincide with four of these. We therefore identify them with 48-51 there. Summing all these things up, we obtain the following table:

I,1-7	53 -56	V,1-11	74'-79'	VIII,1-3	91'-92'
II,1-14	56'-63	IX,1-4	80 -81'	C.1-9	93 -101
III,1-3	63'-64'	X,1-9	82 -86	D.1-3	48 -50
XI,1-9	65 -69	VI,1-6	86'-89	s.1	51
IV, 1-9,*	69'-74	VII,1-3,*	89'-91		

Now we obtain this coded sequence:

53	93 ,	62 Sp'	69	0 C ,	96	85	51 h ,	48 S
54	94 ,	63 Sp	70	62 Sp',	97	86	56 C',	95
55	95 ,	42 h	71	63 Sp,	98	87	98 h ,	38 h-
56	48 S ,	48	72	38 h-,	95 h	88	42 h ,	50 S
57	59 Sp,	38 h	73	48 S ,	56 C'	89	65 C ,	48
58	42 h ,	60 Sp'	74	- ,	99	90	50 S ,	80 C
59	- ,	R	75	62 Sp',	100	91	- ,	48
60	51 S ,	-	76	63 Sp,	95	92	49 S ,	80 C
61	R ,	51 h	77	42 h ,	49 S	93	63_0	
62	45 h-,	- Cc'	78	95 ,	98 h	94	69_0	
63	- C',	95	79	38 h-,	50 S	95	$(\bar{q}-1)_0$	
64	42 h ,	50 S	80	49 ,	50 h-	96	74_0	
65	101 ,	95 h-	81	42 h ,	65 Cc	97	86_0	
66	50 h ,	41 S	82	49 ,	50 h	98	$(N-1)_0$	
67	95 ,	101 h-	83	51 Sp',	51	99	89_0	
68	42 h-,	44 S	84	R ,	51 Sp	100	91_0	
						101	$(\bar{p}-1)_0$	

The state of 0-52 of 10.5 must be exactly as described at the end of 10.7.

The durations may be estimated as follows:

I: 275 μ, II: 485 μ , III: 125 μ , IV: 350 μ , V: 425 μ , VI: 225 μ , VII: 125 μ , VIII: 125 μ , IX: 150 μ , X: 330 μ , XI: 350 μ.

Total: $\mathrm{I} + [\mathrm{II} \text{ or } \mathrm{II} \times 2 \text{ or } \mathrm{II} \times (i_o+1)] + [\mathrm{III} \text{ or } (\mathrm{IV} + \mathrm{VI}) \text{ or } \{(\mathrm{IV} + \mathrm{V} +$

$+ (\mathrm{VII} \text{ or } \mathrm{VIII}) \times (i_o-1) + \mathrm{IX} \times i_o + \mathrm{X} \times (i_o-1)\}] + \mathrm{XI}$

(the two first alternatives in each bracket [ ] refer to two possibilities which can be described by $i_o = 1$ and $i_o = 0$)

$$\text{maximum} = (275 + 485\,(\bar{i}_o+1) + (775 + 125\,(i_o-1) + 150\,i_o + 330\,(i_o-1) + 350)\mu =$$

$$= (1{,}090\,i_o + 1{,}430)\,\mu.$$

$$\text{maximum} = (1{,}090\,{}^2\log\,(N\text{-}M\text{-}2) + 3610)\,\mu \approx (1.1\,{}^2\log\,(N\text{-}M\text{-}2) + 3.6)\text{ m.}$$

Hence the complete interpolation procedure (10.5 plus 10.8) requires

$$(.9\,M^2 - .4\,M + 1.1\,{}^2\log\,(N\text{-}M\text{-}2) + 3)\text{ m.}$$

10.9 To conclude, we take up a variant of the group of Problems 12-13, which illustrates the way in which minor changes in the organization of a problem can be effected ex post, i.e., after the problem has been coded on a different basis. (In this connection, cf. also the discussion of 8.9.)

The description of the function under consideration in Problems 12 and 13.b, i.e., the two sequences $p_1,\ldots,p_M$ and $x_1,\ldots,x_M$ on one hand, and $q_1,\ldots,q_N$ and $y_1,\ldots,y_N$ on the other, were stored in two separate systems of consecutive memory locations: $p,\ldots,p+M-1$ and $q,\ldots,q+M-1$ on one hand, and $\bar{p},\ldots,\bar{p}+N-1$ and $\bar{q},\ldots,\bar{q}+N-1$ on the other. (These are the two storage areas A and B of those two problems.) For Problem 13.a the situation is different, since here $q_1,\ldots,q_N$ alone are stored (at the locations $\bar{p},\ldots,\bar{p}+N-1$, storage area A).

Assume now, that these data are stored together with the p_i, x_1, or the q_i, y_1, alternating. I.e., at the locations $p,\ldots,p+2M-1$, or $\bar{p},\ldots,\bar{p}+2N-1$, with p_i, x_i at $p+2i-2$, $p+2i-1$, or q_1, y_i at $\bar{p}+2i-2$, $\bar{p}+2i-1$.

Since this variant cannot arise for Problem 13.a (cf. above), and since its discussion for Problem 13.b comprises the same for Problem 12 (because the coding of the former requires combining with the coding of the latter, cf. 10.8), therefore we will discuss it for Problem 13.b.

We state accordingly:

PROBLEM 13.c.

Same as Problem 13 in its form 13.b, but the quantities $y_1,\ldots,y_N$ and $q_1,\ldots,q_N$ are stored in one system of 2N consecutive memory locations: $\bar{p}, \bar{p}+1,\ldots,\bar{p}+2N-1$ in this order: $q_1, y_1,\ldots,q_N, y_N$. ----

We must analyze the coded sequences 0-52 of 10.5 and 53-101 of 10.8 which correspond to Problems 12 and 13.b, and determine what changes are necessitated by the new formulation, i.e., by Problem 13.c.

Let us consider the code of 10.8, i.e., of Problem 13.b, first. The changes here are due to two causes: First, y_i is now stored at $\bar{p}+2i-1$ instead of $\bar{q}+i-1$; second, since the code of 10.5, i.e., of Problem 12, will also undergo changes, the substitutions into it will change.

The first group is most simply handled in this way: Assume that $\bar{p}$ is odd. Replace the position marks $(B.i)_0$, which should be $(\bar{p}+2i-1)_0$ instead of $(\bar{q}+i-1)_0$, by their half values: $(\frac{1}{2}(\bar{p}+2i-1))_0 = (\frac{\bar{p}+1}{2}+i-1)_0$. We note, for later use, that this has the effect that D.3 will contain immediately before XI $(\frac{1}{2}(\bar{p}+2\bar{k}-1))_0$ instead of $(B.\bar{k})_0$.

Returning to the general question: we must change the coding of all those places, where such a position mark is used to obtain the corresponding y_i. The values of these position marks must be doubled before they are used in this way. Apart from this, however, we must only see to it that $\bar{q}$ is replaced by $\frac{\bar{p}+1}{2}$. This affects C.3 (of 10.8), i.e., 95.

The use of position marks in the above sense occurs only in II, when y_k and y_{k+M} are obtained. This takes place at II,6 and II,9. However, the position marks in question originate in the substitutions II,2 and II,5, and it is therefore at these places that the changes have to apply.

We now formulate an adequate coding to replace II.2-4, so as to give II,2 and II,5 the desired effect. Note that at this point D.1 contains $(\frac{1}{2}(\bar{p}+2k_i^o-1)) = (\frac{\bar{p}+1}{2}+k_i^o-1)_0$ instead of $(\bar{q}+k_i^o-1)_0 = (B.k_i^o)_0$. Hence after II,1 the accumulator contains $(\frac{1}{2}(\bar{p}+2k_i^o-1))_0$.

II,1.1		L		Ac	$(\bar{p}+2k_i^o-1)_o$
2	II,6	Sp		II,6	$\bar{p}+2k_i^o-1$
3	38	h		Ac	$(\bar{p}+2k_i^o+M-2)_o$
4	38	h		Ac	$(\bar{p}+2k_i^o+2M-3)_o$
5	42	h		Ac	$(\bar{p}+2k_i^o+2M-2)_o$
6	42	h		Ac	$(\bar{p}+2k_i^o+2M-1)_o$

Thus we must replace the three orders II,2-4 by the six orders II,1.1-6. Hence the third and second remarks of 8.9 apply: We replace II,2-3 by II,1.1-2, then II,4 by

II,1.2,1	II,1.3	C	

and let II,1.3-6 be followed by

II,1.7	II,5	C	

Then II,1.3-7 can be placed at the end of the entire coded sequence. We will give a final enumeration that expresses these changes after we have performed all the changes that are necessary.

The second group must be considered in conjunction with the code of 10.5, i.e., of Problem 12. In the code of 10.8 only XI is involved in the operations of this group. In the code of 10.5 the situation is as follows: Replace the position marks A.i, B.i (of 10.5), which should be $(p+2i-2)_o$, $(p+2i-1)_o$, i.e., $(\bar{p}+2\bar{k}+2i-4)_o$, $(\bar{p}+2\bar{k}+2i-3)_o$, instead of $(p+i-1)_o$, $(q+i-1)_o$, i.e., $(\bar{p}+\bar{k}+i-2)_o$, $(\bar{q}+\bar{k}+i-2)_o$, as nearly by their half values as possible. Since $\bar{p}$ is odd, we will use the half value of the second expression: $(\frac{1}{2}(\bar{p}+2\bar{k}+2i-3))_o = (\frac{\bar{p}+1}{2}+\bar{k}+i-2)_o$. Then we must change the coding of all those places where such position marks are used to obtain the corresponding p_i and y_i: The values of these position marks must be doubled before they are used for a y_1, and then decreased by 1 before they are used for a p_1. We must also take care that the storage locations, which supply the modified $(A.i)_o$, $(B.i)_o$ position marks, are properly supplied themselves. Apart from these, however, we must only see to it that XI (of 10.8) produces $(\frac{1}{2}(\bar{p}+2k-1))_o = (\frac{\bar{p}+1}{2}+\bar{k}+i-2)_o\big|_{i=1}$ instead of $(\bar{p}+\bar{k}-1)_o = (\bar{p}+\bar{k}+i-2)_o\big|_{i=1}$. Since p, q, or $\bar{p}$, $\bar{q}$, no longer appear independently, it is not necessary to produce $(\bar{q}-\bar{p}-1)_o$ in XI.

The use of position marks in the above sense occurs only in III, V, and VII (of 10.5), when p_1, P_1, and x_i, x_{i+h} are obtained. These take place at III,3, V,2, and VII,14,15,21. However, the position marks in question originate in the substitutions III,2, V,1, and VII,8,9,13, and it is therefore at these places that the changes have to apply.

The position marks $(A.i)_0$, $(B.i)_0$ are supplied from E.2, which in turn is supplied by II and IX (of 10.5). II will now supply to E.2 $(\frac{1}{2}(\bar{p}+2\bar{k}+1))_0 = (\frac{\bar{p}+1}{2} + \bar{k} + i-2)_0|_{i=2}$ instead of $(p+1)_0 = (\bar{p}+\bar{k})_0 = (\bar{p}+\bar{k}+i-2)_0|_{i=2}$. The function of IX is actually performed by VIII, and it will maintain in E.2 $(\frac{1}{2}(\bar{p}+2\bar{k}+2i-1))_0 = (\frac{\bar{p}+1}{2} + \bar{k} + i-1)_0$ instead of $(p+i(1))_0 = (\bar{p}+\bar{k}+i-1)_0$. If we keep these facts in mind when modifying III, V, VII, then we need not change II, IX (i.e., VIII).

We now give adequate codings, to insert before III,2 and before V,1, and also to replace VII,7 before VII,8,9, and to replace VII,10-12 before VII,13, so as to give III,2, V,1, and VII,8,9,13 the desired effect. Note that at these points, as we observed above, D.4 (i.e., 41, substituted from XI of 10.8) contains $(\frac{1}{2}(\bar{p}+2\bar{k}-1))_0$, and E.2 contains $(\frac{1}{2}(\bar{p}+2\bar{k}+1))_0$ and $(\frac{1}{2}(\bar{p}+2\bar{k}+2i-1))_0$, respectively. Hence before III,2, V,1, and VII,7, the accumulator contains $(\frac{1}{2}(\bar{p}+2\bar{k}-1))_0$, $(\frac{1}{2}(\bar{p}+2\bar{k}+2i-1))_0$ and $(\frac{1}{2}(\bar{p}+2\bar{k}+2i-1))_0$, respectively.

III,1.1		L	Ac	$(\bar{p}+2\bar{k}-1)_0$
2	D.5	h-	Ac	$(\bar{p}+2\bar{k}-2)_0$
V,0.1		L	Ac	$(\bar{p}+2\bar{k}+2i-1)_0$
2	D.5	h-	Ac	$(\bar{p}+2\bar{k}+2i-2)_0$
VII,6.1	D.5	h-	Ac	$(\frac{1}{2}(\bar{p}+2\bar{k}+2i-1)-1)_0$
2		L	Ac	$(\bar{p}+2\bar{k}+2i-3)_0$
VII,9.1	E.2		Ac	$(\frac{1}{2}(\bar{p}+2\bar{k}+2i-1))_0$
2	D.1	h	Ac	$(\frac{1}{2}(\bar{p}+2\bar{k}-2i-1)+M-1)_0$
3	E.1	h-	Ac	$(\frac{1}{2}(\bar{p}+2\bar{k}+2i-1)+h-1)_0$
4		L	Ac	$(\bar{p}+2\bar{k}+2i+2h-3)_0$

Thus we must insert the orders III,1.1-2 before III,2; insert the orders V,0.1-2 before V,1; replace the order VII,7 by the two orders VII,6.1-2; and replace the three orders VII,10-12 by the four orders VII,9.1-4. Hence the third and second remarks of 8.9 apply again: We replace III,1 by

III,0	III,1.0	C	

and let III,1.0 coincide with III,1 and be followed by III,1.1-2 and then by

III,1.3	III,2	C	

Next we replace V,1 by

V,0	V,0.1	C	

and let V,0.1 be as specified above, and be followed by V,0.2, then by V,1, and finally by

V,1.1	V,2	C	

Finally we replace VII,7 and VII,10-12 by VII,6.1-2 and VII,9.1-4. It is best to effect this as a replacement of the entire piece VII,7-12 by VII,6.1-2, VII,8-9, VII,9.1-4 (a total of six orders to be replaced by a total of eight orders). This means that we replace VII,7-11 by VII,6.1-2, VII,8-9, VII,9.1, then we replace VII,12 by

VII,9.1.1	VII,9.2	C	

and let VII,9.2-4 be followed by

VII,9.2.5	VII,13	C	

Then III,1.0-3; V,0.1-2, V,1, V,1.1; VII,9.2-5 can be placed at the end of the entire coded sequence. We will give a final enumeration that expresses these changes after we have performed all the changes that are necessary.

Returning to the second group in the coding of 10.8, we note that it affects only XI (of 10.8). We saw above that XI must produce $(\frac{1}{2}(\bar{p}+2\bar{k}-1))_0$ and substitute it into 41, and that this is all that has to be done there. However, we noted before that D.3 before XI contains precisely $(\frac{1}{2}(\bar{p}+2\bar{k}-1))_0$. Hence XI,1-9 (i.e., XI in its entirety) may be replaced by

XI,0.1	D.3			Ac	$(\frac{1}{2}(\bar{p}+2\bar{k}-1))_0$
2	41	S		41	$(\frac{1}{2}(\bar{p}+2\bar{k}-1))_0$
3	e'	C			

Hence we replace XI,1-3 by XI,0.1-3, and since operations of the code of 10.8 end at this point, XI,4-9 may be left empty.

To conclude, we observe that C.3 must contain $(\frac{\bar{p}-1}{2})_0 = (\frac{\bar{p}+1}{2} - 1)_0$ instead of $(\bar{q}-1)_0$, and that neither $\bar{p}$ nor $\bar{q}$ are needed in any other form, so that C.9 may be left empty.

We are now in possession of a complete list of all changes of both codes 0-52 of 10.5 and 53-101 of 10.8. We tabulate the omissions and modifications first, and the additions afterwards.

Omissions and Modifications:

	From	To
(10.8) :	II,2-3	II,1.1-2
	II,4	II,1.2.1
(10.5) :	III,1	III,0
	V,1	V,0
	VII,7-8	VII,6.1-2
	VII,9-10	VII,8-9
	VII,11	VII,9.1
	VII,12	VII,9.1.1

Omissions and Modifications (cont.):

	From	To
(10.8) :	XI,1-3	XI,0.1-3
	XI,4-9	Empty
	C.3	$(\frac{p-1}{2})_0$
	C.9	Empty

Additions:

(10.8) :	II,1.3-7
(10.5) :	III,1.0-3
(10.5) :	V,0.1-2 V,1 V,1.1
(10.5) :	VII,9.2-5

We can use five of the six empty fields XI,4-9 (from 10.8) to accommodate II,1.3-7 (from 10.8) -- say XI,4-8. C.9 is the last word of the code: 101, hence we can place III,1.0-3; V,0.1-2, V,1, V,1.1, VII,9.2-5 (from 10.5) in a sequence that begins there. They will then occupy these positions:

III,1.0-3	101-102'	V,1	104	VII,2-5	105-106'
V,0.1-2	103-103'	V,1.1	104'		

(All these are from 10.5.)

Now we can formulate the final form of all changes in the coded sequences 0-52 and 53-101:

5	101 C ,		58	66 C' ,		95	$(\frac{p-1}{2})_0$	
10	103 C ,		65	50 ,	41 S	101	41 ,	L
14		42 h-	66	e'C ,	38 h	102	42 h-,	5 C'
15	L ,	18 Sp'	67	38 h ,	42 h	103	L ,	42 h-
16	21 Sp',	49	68	42 h ,	58 C'	104	10 Sp',	10 C'
17	105 C ,		69	-- ,		105	38 h ,	48 h-
57	L ,	59 Sp				106	L ,	17 C'

The following durations are affected:

(10.5) :	III: + 130 μ
	V: + 130 μ
	VII: + 130 μ
(10.8) :	II: + 180 μ
	IX: - 225 μ

Total: (10.5) : $(130 + 130\ (M\text{-}1) + 130\ \frac{M(M\text{-}1)}{2})\ \mu =$

$= (65\ M^2 + 65\ M)\ \mu \approx (.07\ M^2 + .07\ M)\ m.$

Total: (10.8) : $(180\ (i_o + 1) - 225\ i_o)\ \mu = (-\ 45\ i_o + 180)\ \mu$

maximum $\approx .2$ m.

Hence there is no relevant change in the estimate at the end of 10.8.

11.0 CODING OF SOME COMBINATORIAL (SORTING) PROBLEMS

11.1 In this chapter we consider problems of a combinatorial, and not analytical character. This means, that the properly calculational (arithmetical) parts of the procedure will be very simple (indeed almost absent), and the essential operations will be of a logical character. Such problems are of a not inconsiderable practical importance, since they include the category usually referred to as *sorting problems*. They are, furthermore, of a conceptual interest, for the following reason.

Any computing machine is organized around three vital organs: The memory, the logical control, and the arithmetical organ. The last mentioned organ is an adder-subtractor-multiplier-divider, and the multiplier is usually that aspect which primarily controls its speed (cf. the discussions of Part I.) The efficiency of the machine on all analytical problems is clearly dependent to a large extent on the efficiency of the arithmetical organ, and thus primarily of the multiplier. Now the non-analytical, combinatorial problems, to which we have referred, avoid the multiplier almost completely. (Adding and subtracting is hardly avoidable even in purely logical procedures, since the operations are needed to construct the position marks of memory locations and to effect size comparisons of numbers, by which we have to express our logical alternatives.) Consequently these problems provide tests for the efficiency of the non-arithmetical parts of the machine: The memory and the logical control.

11.2 We will consider two typical sorting problems: Those of *meshing* and of *sorting* proper. There are, of course, many others, and many variants for each problem, but these two should suffice to illustrate the main methodical principles.

In order to formulate our two problems, we need the definitions which follow.

We operate with *sequences* of *complexes*. A complex $X = (x;\ u_1,\dots,u_p)$ consists of p+1 numbers: The *principal number* x and the *subsidiary numbers* $u_1,\dots,u_p$. p is the *order* of the complex. A *sequence* $S = (X^{(1)},\dots,X^{(n)})$ consists of n complexes $X^{(i)} = (x^{(i)};\ u_1^{(i)},\dots,u_p^{(i)})$. n is the *length* of the sequence. Throughout what follows we will never vary p, the (arbitrary but) fixed order of all complexes to be considered. We will, however, deal with sequences of various lengths.

A sequence $S = (X^{(1)},\dots,X^{(n)})$ is a *monotone* if the principal elements $x^{(i)}$ of its complexes $X^{(i)} = (x^{(i)};\ u_1^{(i)},\dots,u_p^{(i)})$ form a monotone, non decreasing sequence:

$$(1) \qquad x^{(1)} \leqq x^{(2)} \leqq \dots \leqq x^{(n)}$$

A sequence $T = (Y^{(1)},\dots,Y^{(n)})$ is a *permutation* of another sequence $S = (X^{(1)},\dots,X^{(n)})$ of the same length, if for a suitable permutation $i \rightarrow i'$ of the $i = 1,\dots,n$

$$(2) \qquad Y^{(i)} = X^{(i')} \qquad (i = 1,\dots,n).$$

Given two sequences $S = (X^{(1)},\ldots,X^{(n)})$, $T = (Y^{(1)},\ldots,Y^{(m)})$ (of not necessarily equal lengths n, m), their *sum* is the sequence $[S, T] = (X^{(1)},\ldots,X^{(n)}, Y^{(1)},\ldots,Y^{(m)})$ (of length n + m).

We can now state our two problems:

PROBLEM 14. (Meshing)

Two monotone sequences *S, T*, of lengths n, m, respectively, are stored at two systems of n(p+1), m(p+1) consecutive memory locations respectively: s, s+1,..., s+n(p+1)-1 and t, t+1,..., t+m(p+1)-1. An accomodation for a sequence of length n + m, consisting of (n + m)(p + 1) consecutive memory locations, is available: r, r+1,..., r+(n+m)(p+1)-1. The constants of the problem are p, n, m, s, t, r, to be stored at six given memory locations. It is desired to find a monotone permutation *R* of the sum *[S, T]*, and to place it at the locations r, r+1,..., r+(n+m)(p+1)-1. ----

PROBLEM 15. (Sorting)

A sequence *S* (subject to no requirements of monotony whatever) is stored at n(p+1) consecutive memory locations: s, s+1,..., s+n(p+1)-1. The constants of the problem are s, n, p, to be stored at three given memory locations. It is desired to find a monotone permutation S^* of *S*, and to replace *S* by it.

11.3 We consider first Problem 14, which is the simpler one, and whose solution can be conveniently used in solving Problem 15.

We have $S = (X^{(1)},\ldots, X^{(n)})$, $T = (Y^{(1)},\ldots, Y^{(m)})$ with $X^{(i)} = (x^{(i)}; u_1^{(i)},\ldots, u_p^{(i)})$, $Y^{(j)} = (y^{(j)}; v_1^{(j)},\ldots, v_p^{(j)})$, and we wish to form $R = (Z^{(1)},\ldots, Z^{(n+m)})$ with $Z^{(k)} = (z^{(k)}; w_1^{(k)},\ldots,w_p^{(k)})$, which is a monotone permutation of *[S, T]*.

Forming *R* can be viewed as an inductive process: If $Z^{(1)},\ldots,Z^{(k-1)}$ for a k=1,...,n+m, have already been formed, the inductive step consists of forming $Z^{(k)}$. It is preferable to allow k = n+m+1, too, in the sense that when this value of k is reached, the process is terminated. Assume, therefore, that we have formed $Z^{(1)},\ldots,Z^{(k-1)}$ for a k = 1,...,n+m+1. It is clear, that we may assume in addition, that these $Z^{(1)},\ldots,Z^{(k-1)}$ are the $X^{(1)},\ldots,X^{(i_k)}$ together with the $Y^{(1)},\ldots,Y^{(j_k)}$, for two suitable i_k, j_k with

(1) $$i_k + j_k = k-1 \qquad i_k = 1,\ldots, n, \quad j_k = 1,\ldots, m.$$

Then we have the following possibilities:

(a) $i_k = n$, $j_k = m$: By (1) k = n+m+1, hence, as we observed above, *R* is fully formed, and nothing remains to be done.

(b) Not $i_k = n$, $j_k = m$: By (1) $k \leq n+m$, hence, as we observed above, *R* is not fully formed, and $Z^{(k)}$ must be formed next. Clearly, we may choose $Z^{(k)}$ as one of $X^{(i_k+1)}$ and $Y^{(j_k+1)}$, we must decide which. We introduce a quantity ω_k, so that $\omega_k = 0$ for $Z^{(k)} = X^{(i_k+1)}$ and $\omega_k = 1$ for $Z^{(k)} = Y^{(j_k+1)}$.

$\omega_k = 0$ implies $i_{k+1} = i_k + 1$, $j_{k+1} = j_k$; $\omega_k = 1$ implies $i_{k+1} = i_k$, $j_{k+1} = j_k + 1$. ω_k is determined according to the following rules:

(b.1) $i_k = n$, hence $j_k \neq m$: S is exhausted, hence $\omega_k = 1$.

(b.2) $j_k = m$, hence $i_k \neq n$: T is exhausted, hence $\omega_k = 0$.

(b.3) $i_k \neq n$, $j_k \neq m$: Neither S nor T is exhausted. We must distinguish further:

(b.3.1) $x^{(i_k+1)} < y^{(j_k+1)}$: Necessarily $\omega_k = 0$.

(b.3.2) $x^{(i_k+1)} > y^{(j_k+1)}$: Necessarily $\omega_k = 1$.

(b.3.3) $x^{(i_k+1)} = y^{(j_k+1)}$: Both $\omega_k = 0$ and $\omega_k = 1$ are acceptable. We choose $\omega_k = 0$.

As pointed out above, (a), (b) describes an inductive process which runs over $k = 1, \ldots, n+m+1$. It begins with $k = 1$ and $i_k = j_k = 0$ (owing to (1)), and then progresses from k to $k+1$, forming the necessary i_k, j_k and ω_k as it goes along.

It is convenient to denote the locations s and t, where the storage of the sequences S and T begins, by a common letter and to distinguish them by an index:

(2) $$s^0 = s, \qquad s^1 = t.$$

Correspondingly, let A^0 and A^1 be the storage areas corresponding to the two intervals of locations from s^0 to $s^0 + n(p+1) - 1$ and from s^1 to $s^1 + m(p+1) - 1$. In this way their positions will be $A^0.1, \ldots, n(p+1)$ and $A^1.1, \ldots, m(p+1)$, where $A^0.a$ and $A^1.b$ correspond to $s^0 + a - 1$ and $s^1 + b - 1$, and store $x^{(i)}$ for $a = (i-1)(p+1)+1$, $u_q^{(i)}$ for $a = (i-1)(p+1)+q+1$, $y^{(j)}$ for $b = (j-1)(p+1)+1$, $v_q^{(j)}$ for $b = (j-1)(p+1)+q+1$, respectively. Next, let B be the storage area intended for the sequence R, corresponding to the interval of locations from r to $r+(n+m)(p+1)-1$. In this way its positions will be $B.1, \ldots, (n+m)(p+1)$, where B.c corresponds to $r+c-1$, and is intended to store $z^{(k)}$ for $c = (k-1)(p+1)+1$, $w_q^{(k)}$ for $c = (k-1)(p+1)+q+1$, respectively. As in several previous problems, the positions of A^0, A^1, and B need not be shown in the final enumeration of the coded sequence.

Further storage capacities are needed as follows: The given data (the constants) of the problem, s^0, s^1, r, n, m, p, will be stored in the storage area C. (It will be convenient to store them as $(s^0)_0$, $(s^1)_0$, r_0, $(n(p+1))_0$, $(m(p+1))_0$, $(p+1)_0$. We will also store 1_0 in C.) Next the induction indices k, q' will have to be stored. It is convenient to store i_k, j_k, too, along with k, while it turns out to be unnecessary to form and to store ω_k explicitly. These indices i, j, k, q' are all relevant as position marks, and it will prove convenient to store in their place $(s^0 + i(p+1))_0$, $(s^1 + j(p+1))_0$, $(r + (k-1)(p+1))_0$ $(s^\omega + h(p+1) + q' - 1)_0$, $(r + (k-1)(p+1) + q' - 1)_0$ (i.e. $(A^0.i(p+1)+1)_0$, $(A^1.j(p+1)+1)_0$, $(B.(k-1)(p+1)+1)_0$, $(A^\omega.h(p+1) + q')_0$, $(B.(k-1)(p+1) + q')_0$; here $\omega = \omega_k$, and h = i or j for $\omega = 0$ or 1, respectively) Note, that two position marks that correspond to q' are being stored. The three first quantities will be stored in the storage area D, and the two last ones in the storage area E.

We can now draw the flow diagram as shown in Figure 11.1. The actual coding obtains from this without a need for further comments.

The static coding of the boxes I-X follows:

C.1	$(s^0)_0$				
I,1	C.1		Ac	$(s^0)_0$	
2	D.1	S	D.1	$(s^0)_0$	
C.2	$(s^1)_0$				
I,3	C.2		Ac	$(s^1)_0$	
4	D.2	S	D.2	$(s^1)_0$	
C.3	r_0				
I,5	C.3		Ac	$(r)_0$	
6	D.3	S	D.3	$(r)_0$	
(to II,1)					
C.4	$(n(p+1))_0$				
II,1	D.1		Ac	$(s^0+i(p+1))_0$	
2	C.1	h-	Ac	$(i(p+1))_0$	
3	C.4	h-	Ac	$((i-n)(p+1))_0$	
4	IV,1	Cc			
(to III,1)					
C.5	$(m(p+1))_0$				
III,1	D.2		Ac	$(s^1+j(p+1))_0$	
2	C.2	h-	Ac	$(j(p+1))_0$	
3	C.5	h-	Ac	$((j-m)(p+1))_0$	
4	VI,1	Cc			
(to V,1)					
IV,1	D.2		Ac	$(s^1+j(p+1))_0$	
2	C.2	h-	Ac	$(j(p+1))_0$	
3	C.5	h-	Ac	$((j-m)(p+1))_0$	
4	e	Cc			
(to VII,1)					
V,1	D.1		Ac	$(s^0+i(p+1))_0$	
2	V,6	Sp	V,6	$s^0+i(p+1)$	h-
3	D.2		Ac	$(s^1+j(p+1))_0$	
4	V,5	Sp	V,5	$s^1+j(p+1)$	
5	--				
[	$s^1+j(p+1)$]		Ac	y_{j+1}	
6	--	h-			
[	$s^0+i(p+1)$ h-]		Ac	$y_{j+1}-x_{i+1}$	
7	VI,1	Cc			
(to VII,1)					

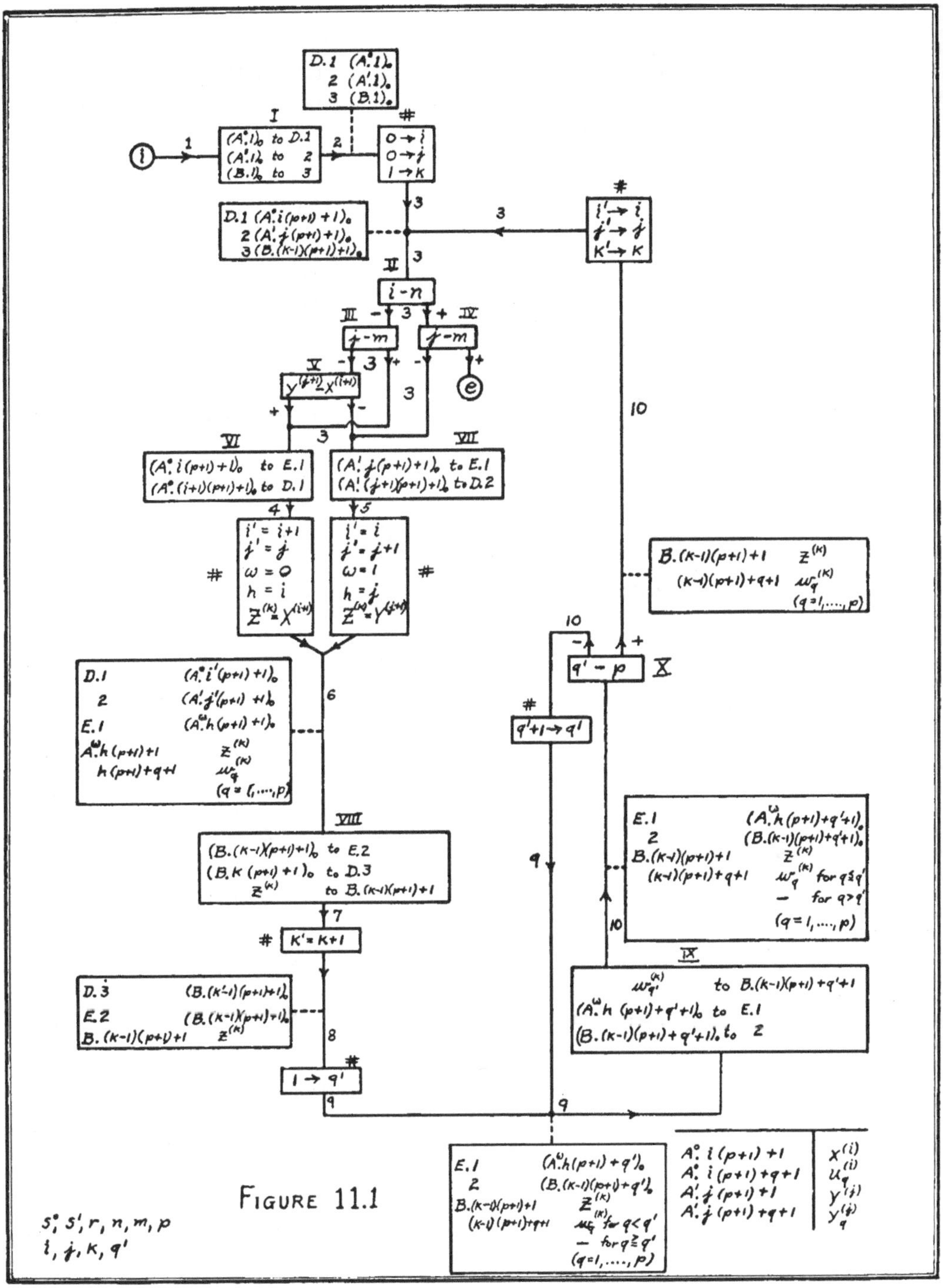
D.1 (A°.1)₀
2 (A'.1)₀
3 (B.1)₀
I
(A°.1)₀ to D.1
(A'.1)₀ to 2
(B.1)₀ to 3
0 → i
0 → j
1 → K
D.1 (A°.i(p+1)+1)₀
2 (A'.j(p+1)+1)₀
3 (B.(K-1)(p+1)+1)₀
i' → i
j' → j
K' → K
II
i - n
III
IV
j - m
j - m
V
Y^(j+1) ≥ X^(i+1)
VI
VII
(A°.i(p+1)+1)₀ to E.1
(A°.(i+1)(p+1)+1)₀ to D.1
(A'.j(p+1)+1)₀ to E.1
(A'.(j+1)(p+1)+1)₀ to D.2
i' = i+1
j' = j
ω = 0
h = i
Z^(K) = X^(i+1)
i' = i
j' = j+1
ω = 1
h = j
Z^(K) = Y^(j+1)
B.(K-1)(p+1)+1 Z^(K)
(K-1)(p+1)+q+1 w_q^(K)
(q = 1, ..., p)
q' - p
X
q'+1 → q'
D.1 (A°.i'(p+1)+1)₀
2 (A'.j'(p+1)+1)₀
E.1 (A^ω.h(p+1)+1)₀
A^ω.h(p+1)+1 Z^(K)
h(p+1)+q+1 w_q^(K)
(q = 1, ..., p)
VIII
(B.(K-1)(p+1)+1)₀ to E.2
(B.K(p+1)+1)₀ to D.3
Z^(K) to B.(K-1)(p+1)+1
E.1 (A^ω.h(p+1)+q'+1)₀
2 (B.(K-1)(p+1)+q'+1)₀
B.(K-1)(p+1)+1 Z^(K)
(K-1)(p+1)+q+1 w_q^(K) for q ≤ q'
— for q > q'
(q = 1, ..., p)
K' = K+1
IX
w_q'^(K) to B.(K-1)(p+1)+q'+1
(A^ω.h(p+1)+q'+1)₀ to E.1
(B.(K-1)(p+1)+q'+1)₀ to 2
D.3 (B.(K-1)(p+1)+1)₀
E.2 (B.(K-1)(p+1)+1)₀
B.(K-1)(p+1)+1 Z^(K)
1 → q'
E.1 (A^ω.h(p+1)+q')₀
2 (B.(K-1)(p+1)+q')₀
B.(K-1)(p+1)+1 Z^(K)
(K-1)(p+1)+q+1 w_q^(K) for q < q'
— for q ≥ q'
(q = 1, ..., p)
A°.i(p+1)+1 X^(i)
A°.i(p+1)+q+1 u_q^(i)
A'.j(p+1)+1 Y^(j)
A'.j(p+1)+q+1 Y_q^(j)
s°, s', r, n, m, p
i, j, K, q'

FIGURE 11.1

C.6	$(p+1)_0$				
VI,1	D.1		Ac	$(s^0+i(p+1))_0$	
2	E.1	S	E.1	$(s^0+i(p+1))_0$	
3	C.6	h	Ac	$(s^0+(i+1)(p+1))_0$	
4	D.1	S	D.1	$(s^0+(i+1)(p+1))_0$	
(to VIII,1)					
VII,1	D.2		Ac	$(s^1+j(p+1))_0$	
2	E.1	S	E.1	$(s^1+j(p+1))_0$	
3	C.6	h	Ac	$(s^1+(j+1)(p+1))_0$	
4	D.2	S	D.2	$(s^1+(j+1)(p+1))_0$	
(to VIII,1)					
VIII,1	D.3		Ac	$(r+(k-1)(p+1))_0$	
2	VIII,9	Sp	VIII,9	$r+(k-1)(p+1)$	S
3	E.2	S	E.2	$(r+(k-1)(p+1))_0$	
4	C.6	h	Ac	$(r+k(p+1))_0$	
5	D.3	S	D.3	$(r+k(p+1))_0$	
6	E.1		Ac	$(s^{\omega}+h(p+1))_0$	
7	VIII,8	Sp	VIII,8	$s^{\omega}+h(p+1)$	
8	--				
[	$s^{\omega}+h(p+1)$	]	Ac	$z^{(k)}$	
9	--	S			
[	$r+(k-1)(p+1)$	S]	B.(k-1)(p+1)+1	$z^{(k)}$	
(to IX,1)					
C.7	1_0				
IX,1	E.2		Ac	$(r+(k-1)(p+1)+q'-1)_0$	
2	C.7	h	Ac	$(r+(k-1)(p+1)+q')_0$	
3	IX,10	Sp	IX,10	$r+(k-1)(p+1)+q'$	S
4	E.2	S	E.2	$(r+(k-1)(p+1)+q')_0$	
5	E.1		Ac	$(s^{\omega}+h(p+1)+q'-1)_0$	
6	C.7	h	Ac	$(s^{\omega}+h(p+1)+q')_0$	
7	IX,9	Sp	IX,9	$s^{\omega}+h(p+1)+q'$	
8	E.1	S	E.1	$(s^{\omega}+h(p+1)+q')_0$	
9	--				
[	$s^{\omega}+h(p+1)+q'$	]	Ac	$w_{q'}^{(k)}$	
10	--	S			
[	$r+(k-1)(p+1)+q'$	S]	B.(k-1)(p+1)+q'+1	$w_{q'}^{(k)}$	
(to X,1)					

X,1	E.2		Ac	$(r+(k-1)(p+1)+q')_0$
2	D.3	h-	Ac	$(q'-p-1)_0$
3	C.7	h	Ac	$(q'-p)_0$
4	II,1	Cc		
(to IX,1)				

The ordering of the boxes is I, II, III, V, VII, VIII, IX, X, and VII, VIII, IX must also be the immediate successors of IV, VI, X, respectively. This necessitates the extra orders

IV,5	VII,1	C
VI,5	VIII,1	C
X,5	IX,1	C

We must now assign C.1-7, D.1-3, E.1-2 their actual values, pair the 59 orders I,1-6, II,1-4, III,1-4, IV,1-5, V,1-7, VI,1-5, VII,1-4, VIII,1-9, IX,1-10, X,1-5 to 30 words and then assign I,I-X,5 their actual values. These are expressed in this table:

I,1-6	0 -2'	VII,1-4	10'-12	IV,1-5	24'-26'
II,1-4	3 -4'	VIII,1-9	12'-16'	VI,1-5	27 -29
III,1-4	5 -6'	IX,1-10	17 -21'	C.1-7	30 -36
V,1-7	7 -10	X,1-5	22 -24	D.1-3	37 -39
				E.1-2	40 -41

Now we obtain this coded sequence:

0	30 , 37 S	14	35 h , 39 S	28	35 h , 37 S
1	31 , 38 S	15	40 , 16 Sp	29	12 C', - -
2	32 , 39 S	16	- , - S	30	$(s^0)_0$
3	37 , 30 h-	17	41 , 36 h	31	$(s^1)_0$
4	33 h-, 24 Cc'	18	21 Sp', 41 S	32	r_0
5	38 , 31 h-	19	40 , 36 h	33	$(n(p+1))_0$
6	34 h-, 27 Cc	20	21 Sp, 40 S	34	$(m(p+1))_0$
7	37 , 9 Sp'	21	- , - S	35	$(p+1)_0$
8	38 , 9 Sp	22	41 , 39 h-	36	1_0
9	- , - h-	23	36 h , 3 Cc	37	-
10	27 Cc, 38	24	17 C , 38	38	-
11	40 S , 35 h	25	31 h-, 34 h-	39	-
12	38 S , 39	26	e Cc, 10 Cc'	40	-
13	16 Sp', 41 S	27	37 , 40 S	41	-

The durations may be estimated as follows:

I: 225 μ, II: 150 μ, III: 150 μ, IV: 200 μ, V: 275 μ, VI: 200 μ, VII: 150 μ, VIII: 350 μ, IX: 375 μ, X: 200 μ.

$$\text{Total: } I + \left\{ \begin{matrix} II + (III \text{ or } (III + V) \text{ or } IV) + (VI \text{ or } VII) + \\ + VIII + (IX + X) \times p \end{matrix} \right\} \times (n + m) + (II + IV)$$

$$\text{maximum} = (225 + (150 + 425 + 200 + 350 + 575 \times p) \times (n + m) + 350)\ \mu =$$
$$= ((575\ p + 1{,}125)\ (n + m) + 575)\ \mu \approx$$
$$\approx .6\ ((p + 2)\ (n + m) + 1)\ m$$

This result will be analyzed further in 11.5.

11.4 We pass now to Problem 15, which will be solved, as indicated above, with the help of the solution of Problem 14.

We have $S = (X^{(1)}, \ldots, X^{(n)})$ with $X^{(i)} = (x^{(i)};\ u_1^{(i)}, \ldots, u_p^{(i)})$, and we wish to form an $S^* = (X^{(1')}, \ldots, X^{(n')})$, which is a monotone permutation of S.

This can be achieved in the following manner: If a certain permutation $S^{*o} = (X^{(1'^o)}, \ldots, X^{(n'^o)})$ of S has been found, which may not yet be fully monotone, but in which at least two consecutive intervals $X^{(k'^o)}, \ldots, X^{(k+i-1'^o)}$ and $X^{(k+i'^o)}, \ldots, X^{(k+i+j-1'^o)}$ are (each one separately) monotone, then we can mesh these (in the sense of Problem 14), and thereby obtain a new permutation $S^{*oo} = X^{(1'^{oo})}, \ldots, X^{(n'^{oo})})$ of S^{*o}, i.e. of S, for which the whole interval $X^{(k'^{oo})}, \ldots, X^{(k+i+j-1'^{oo})}$ is monotone. Thus we can convert two monotone intervals of the lengths i, j into one of the length i + j. The original S is made up of monotone intervals of length 1, since each $X^{(k)}$ (separately) may be viewed as such an interval. Hence we can, beginning with these monotone intervals of length 1, build up monotone intervals of successively increasing length, until the maximum length n is reached, i.e. the permutation of S at hand is monotone as a whole.

This qualitative description can be formalized to an inductive process. This process goes through the successive stages v = 0, 1, 2,... . In the stage v a permutation $S^{*v} = (X^{(1'^v)}, \ldots, X^{(n'^v)})$ of S is at hand, which is made up of consecutive intervals of length 2^v, which are (each one separately) monotone. (Since n may not be divisible by v, the length of the last interval has to be the remainder r_v of the division of n by 2^v, $r_v < 2^v$. If n is divisible by 2^v, we should put $r_v = 0$ and rule this interval to be absent, but we may just as well put $r_v = 2^v$, and still rule it to be the last interval.) Now a sequence of steps of the nature described above will produce a new permutation S^{*v+1} of S^{*v}, i.e. of S, in which the monotone intervals of length 2^v in S^{*v} are pairwise meshed to monotone intervals of length 2^{v+1} in S^{*v+1}. (The end-effect for v + 1 can be described in the same terms as above for v.) This inductive process begins with v = 0 and $S^{*0} = S$. It ends when a v with $2^v \geqq n$ has been reached, the corresponding S^{*v} is the desired (monotone) S^*. (Note, that at this point there is only one interval: It is the final interval of length $r_v \leqq 2^v$ referred to above, in this case with $r_v = n$.) Using the function $\langle z \rangle$, which denotes the smallest integer $\geqq z$, we may therefore say that this process terminates with $v = \langle {}^2\log n \rangle$.

The induction over v = 0, 1, 2,..., $\langle {}^2\log n \rangle$ is, however, only the primary induction. For each v we have $\left\langle \frac{n}{2^v} \right\rangle$ intervals of length 2^v (cf. above), these form $\left\langle \frac{1}{2} \left\langle \frac{n}{2^v} \right\rangle \right\rangle = \left\langle \frac{n}{2^{v+1}} \right\rangle$ pairs, which have to be meshed to $\left\langle \frac{n}{2^{v+1}} \right\rangle$ intervals of length 2^{v+1} (cf. above), in order to effect the inductive step from v to v + 1.

If we enumerate these interval pairs by an index $w = 1, \ldots, \langle \frac{n}{2^{v+1}} \rangle$, then it becomes clear that we are dealing with a secondary induction, of which w is the index.

We can now state the general inductive step with complete rigor:

Consider a $v = 0, 1, 2, \ldots$, and assume that the permutation $S^{*v} = (X^{(1'v)}, \ldots, X^{(n'v)})$ has already been formed. We effect a secondary induction over a $w = 0, 1, 2, \ldots$. Assume that a certain w has already been reached. Then $2^{v+1}w \leqq n$. We now have to mesh two intervals of the length n_w, m_w, i.e. $X^{(i'^0)}$ with $i = 2^{v+1}w+1, \ldots, 2^{v+1}w+n_w$ and $X^{(j'^0)}$ with $j + 2^{v+1}w + n_w + 1, \ldots, 2^{v+1}w + n_w + m_w$. Ordinarily $n_w = m_w = 2^v$, i.e. this is the case when both these intervals can be accomodated in the interval $1, \ldots, n$. Thus the condition for this case is $2^{v+1}(w+1) \leqq n$. Otherwise, i.e. for $2^{v+1}(w+1) > n$, we have two possibilities: First: $n_w = 2^v$ if this interval can be accomodated in the interval $1, \ldots, n$, i.e. if $2^{v+1}w + 2^v \leqq n$. m_w is then chosen so as to exhaust what is left of the interval $1, \ldots, n$, i.e. $m_w = n - (2^{v+1}w + 2^v)$. Thus the condition for this case is $2^{v+1}(w+1) > n \geqq 2^{v+1}w + 2^v$. We may include $2^{v+1}(w+1) = n$, too, since the formulae of the present case coincide then with those of the previous one. Second: Otherwise, i.e. for $2^{v+1}w + 2^v > n$, the second interval is missing $m_w = 0$. n_w is then chosen so as to exhaust what is left of the interval $1, \ldots, n$, i.e. $n_w = n - 2^{v+1}w$. Thus the condition for this case is $2^{v+1}w + 2^v > n$ (and, of course, $\geqq 2^{v+1}w$). We may include $2^{v+1}w + 2^v = n$, too, since the formulae of the present case coincide then with those of the previous one. These rules can be summed up as follows:

$$
(3) \qquad \begin{cases}
n_w = 2^v, & m_w = 2^v \\
 & \text{for } 2^{v+1}(w+1) \leqq n, \\
n_w = 2^v, & m_w = n - (2^{v+1}w + 2^v) \\
 & \text{for } 2^{v+1}(w+1) \geqq n \geqq 2^{v+1}w + 2^v, \\
n_w = n - 2^{v+1}w, & m_w = 0 \\
 & \text{for } 2^{v+1}w + 2^v \geqq n \ (\geqq 2^{v+1}w).
\end{cases}
$$

If $2^{v+1}(w+1) < n$, then the (secondary) induction over w continues to $w + 1$; if $2^{v+1}(w+1) \geqq n$, then the induction over w stops. If the induction over w stops with a $w > 0$, then the (primary) induction over v continues to $v + 1$; if the induction over w stops with $w = 0$, then the induction over v stops. At any rate these meshings form the permutation S^{*v+1}. If the induction over v stops with a certain v, then its S^{*v+1} is the desired S^*.

There remains, finally, the necessity to specify the locations of S^{*v} and S^{*v+1}, which control the meshing processes that lead from S^{*v} to S^{*v+1}. Assume that S^{*v} occupies the interval of locations from $a^{(v)}$ to $a^{(v)} + n(p+1)-1$, and S^{*v+1} similarly the interval of locations from $a^{(v+1)}$ to $a^{(v+1)} + n(p+1)-1$. Then the meshing process which corresponds to a given w meshes the sequences that occupy the intervals from $a^{(v)} + 2^{v+1}w(p+1)$ to $a^{(v)} + (2^{v+1}w + n_w)(p+1)-1$ and from $a^{(v)} + (2^{v+1}w + n_w)(p+1)$ to $a^{(v)} + (2^{v+1}w + n_w + m_w)(p+1)-1$, and places the resulting sequence into the interval from $a^{(v+1)} + 2^{v+1}w(p+1)$ to $a^{(v+1)} + (2^{v+1}w + n_w + m_w)(p+1)-1$. The meshing process is that one of Problem 14.

We distinguish the constants of that problem by bars. The above specifications of the locations that are involved can then be expressed as follows:

$$(4)\quad \begin{cases} \bar{s}^0 = a^{(v)} + 2^{v+1}w \quad (p+1), \\ \bar{s}^1 = a^{(v)} + (2^{v+1}w + n_w)(p+1), \\ \bar{r} = a^{(v+1)} + 2^{v+1}w \quad (p+1), \\ \bar{n} = n_w, \\ \bar{m} = m_w, \\ \bar{p} = p \end{cases}$$

Clearly the interval from $a^{(v)}$ to $a^{(v)}+n(p+1)-1$ and the interval from $a^{(v+1)}$ to $a^{(v+1)}+n(p+1)-1$ must have no elements in common.

We need such an $a^{(v)}$, i.e. an interval from $a^{(v)}$ to $a^{(v)}+n(p+1)-1$, for every v for which S^{*v} is being formed. It is, however, sufficient to have two such intervals which have no elements in common, say from a to $a+n(p+1)-1$ and from b to $b+n(p+1)-1$. We can then let $a^{(v)}$ alternate between the two values a and b, i.e. we define

$$(5)\quad a^{(v)} \begin{cases} = a \text{ for } v \text{ even} \\ = b \text{ for } v \text{ odd} \end{cases}$$

Hence $a^{(0)} = a$, but $S^{*0} = S$, hence the statement of Problem 15 requires $a^{(0)} = s$. Consequently

$$(6)\quad s = a.$$

Thus a is the s of Problem 14, and b is any such location such that the interval from a to $a+n(p+1)-1$ and the interval from b to $b+n(p+1)-1$ have no elements in common.

We want the final S^* to occupy the location of the original S, i.e. we want $a^{(v+1)} = a^{(0)} = a$ for the final v with $S^{*v+1} = S^*$. I.e. this v+1 should be even. We saw further above, that the induction over v can stop when the induction over w stops with w=0, i.e. when $2^{v+1} \geqq n$. This v+1 may be odd; let us therefore agree to perform one more, seemingly superfluous, step from v to v+1. I.e. we will terminate the induction over v when

$$(7)\quad 2^{v+1} \geqq n \text{ and } v+1 \text{ even, i.e. } a^{(v+1)} = a.$$

Hence this $v+1$ is the smallest even λ with $2^\lambda \geqq n$, i.e. $2\lambda'$ for the smallest integer λ' with $2^{2\lambda'} \geqq n$; i.e. $4^{\lambda'} \geqq n$. This means that

$$(8)\quad v+1 = 2\,\langle {}^4\log n \rangle .$$

In the coded sequence that we will develop for our problem, the coded sequence that solves Problem 14 will occur as a part, as we have already pointed out. This is the coded sequence 0-43 in 11.3. We propose to use it now, and to adjust the coded sequence that we are forming accordingly.

This is analogous to what we did in 10.6 and 10.7 for the Problems 13.a and 13.b: There the coded sequence formed in 10.5 for Problem 12 was used as part of the coded sequences of the two former problems. There is, however, this difference: In 10.6, 10.7 the subsidiary sequence (of Problem 12) was attached to the end of the main sequences (of Problems 13.a, b), while now the subsidiary sequence (of Problem 14) will occur in the interior of the main sequence (of Problem 15) -- indeed it is a part of the inductive step in the double induction over v, w. The constants of Problem 14 have already been dealt with: They must be substituted according to (4) above.

In assigning letters to the various storage areas to be used, it must be remembered, just as at the corresponding points in 10.6, 10.7, that the coded sequence that we are now developing is to be used in conjunction with (i.e. as an extension of) the coded sequence of 11.3. It is therefore again necessary to classify the storage areas required by the latter: We have the storage areas C-E, which are incorporated in the final enumeration (they are 30-41 of the 0-41 of 11.1); and A^0, A^1 and B (i.e. $\bar{s}^0, \ldots, \bar{s}^0 + \bar{n}(\bar{p}+1)-1$; $\bar{s}^1, \ldots, \bar{s}^1 + \bar{m}(\bar{p}+1)-1$ and $\bar{r}, \ldots, \bar{r} + (\bar{n}+\bar{m})(\bar{p}+1)-1$), which will be part of our present A, B (i.e. of $a, \ldots, a + n(p+1)-1$ and $b, \ldots, b + n(p+1)-1$, they will be $a^{(v)} + 2^{v+1}w(p+1), \ldots, a^{(v)} + (2^{v+1}w + n_w)(p+1)-1$; $a^{(v)} + (2^{v+1}w + n_w)(p+1), \ldots, a^{(v)} + (2^{v+1}w + n_w + m_w)(p+1)-1$, and $a^{(v+1)} + 2^{v+1}w(p+1), \ldots, a^{(v+1)} + (2^{v+1}w + n_w + m_w)(p+1)-1$, cf. (4) above). Therefore we can, in assigning letters to the various storage areas in the present coding, disregard those of 11.3.

Let A and B be the storage areas corresponding to the intervals from a to $a+n(p+1)-1$ and from b to $b+n(p+1)-1$. In this way their positions will be A.1,...,n(p+1) and B.1,...,n(p+1), where A.i and B.i correspond to $a+i-1$ and $b+i-1$. A will store S at the beginning and S^* at the end of the procedure. B may be at any available place, and we assume it to be irrelevantly occupied. In the course of the procedure A and B are the alternate values of $A^{(v)}$ (for $a^{(v)} = a$ and b, i.e. for v even and odd, respectively), and $S^{(v)}$ is stored at $A^{(v)}$ while $S^{(v)}$ or $S^{(v+1)}$ is being formed. All these storages are arranged like the storage of S at A^0 in 11.3.

The given data of the problem, a, b, n, p, will be stored in the storage area C, with the exception of p. (It will be convenient to store them as a_0, b_0, $n(p+1)_0$. A 0 will also be stored in C.) p is also needed in the coded sequence of 11.3, and it is the only one of the constants of that problem which is independent of the induction indices v, w. (Cf. (4) above.) Hence its storage in that sequence, at 35 (as $(p+1)_0$) is adequate for our present purpose. The other constants of that problem depend on v, w (cf. above), hence their locations, 30-34, must be left empty (or, rather, irrelevantly occupied), when our present coded sequence begins to operate, and they must be appropriately substituted by its operation.

The induction index v will be stored in D. (It will be convenient to form $(2^{v+1}(p+1))_0$, and to store $(a^{(v)})_0, (a^{(v+1)})_0$ along with it.) The induction index w is part of the expressions which have to be placed at 30 and 32 ($(\bar{s}^0)_0$ and $(\bar{r})_0$, i.e. by (4) $(a^{(v)}+2^{v+1}w\,(p+1))$ and $(a^{(v+1)}+2^{v+1}w\,(p+1))_0$), hence it requires no other storage.

We can now draw the flow diagram, as shown in Figure 11.2. The actual coding obtains from this quite directly, in one instance (at the beginning of box VI) the accumulator is used as storage between two boxes.

The static coding of the boxes I-XIV follows.

C.1	a_0			
2	b_0			
I,1	35		Ac	$(p+1)_0$
2	D.1	S	D.1	$(p+1)_0$
3	C.1		Ac	a_0
4	D.2	S	D.2	a_0
5	C.2		Ac	b_0
6	D.3	S	D.3	b_0
(to II,1)				
II,1	D.2		Ac	$(a^{(v)})_0$
2	30	S	30	$(a^{(v)})_0$
3	D.3		Ac	$(a^{(v+1)})_0$
4	32	S	32	$(a^{(v+1)})_0$
(to III,1)				
III,1	30		Ac	$(a^{(v)}+2^{v+1}w\,(p+1))_0$
2	D.2	h-	Ac	$(2^{v+1}w\,(p+1))_0$
3	D.1	h	Ac	$((2^{v+1}w + 2^{v})(p+1))_0$
C.3	$(n(p+1))_0$			
III,4	C.3	h-	Ac	$((2^{v+1}w+2^{v}-n)(p+1))_0$
5	IV,1	Cc		
(to V,1)				
IV,1	C.3		Ac	$(n(p+1))_0$
2	D.2	h	Ac	$(a^{(v)}+n(p+1))_0$
3	31	S	31	$(a^{(v)}+n(p+1))_0$
4	30	h-	Ac	$((n-2^{v+1}w)(p+1))_0$
5	33	S	33	$((n-2^{v+1}w)(p+1))_0$
C.4	0			
IV,6	C.4		Ac	0
7	34	S	34	0
(to IX,1)				

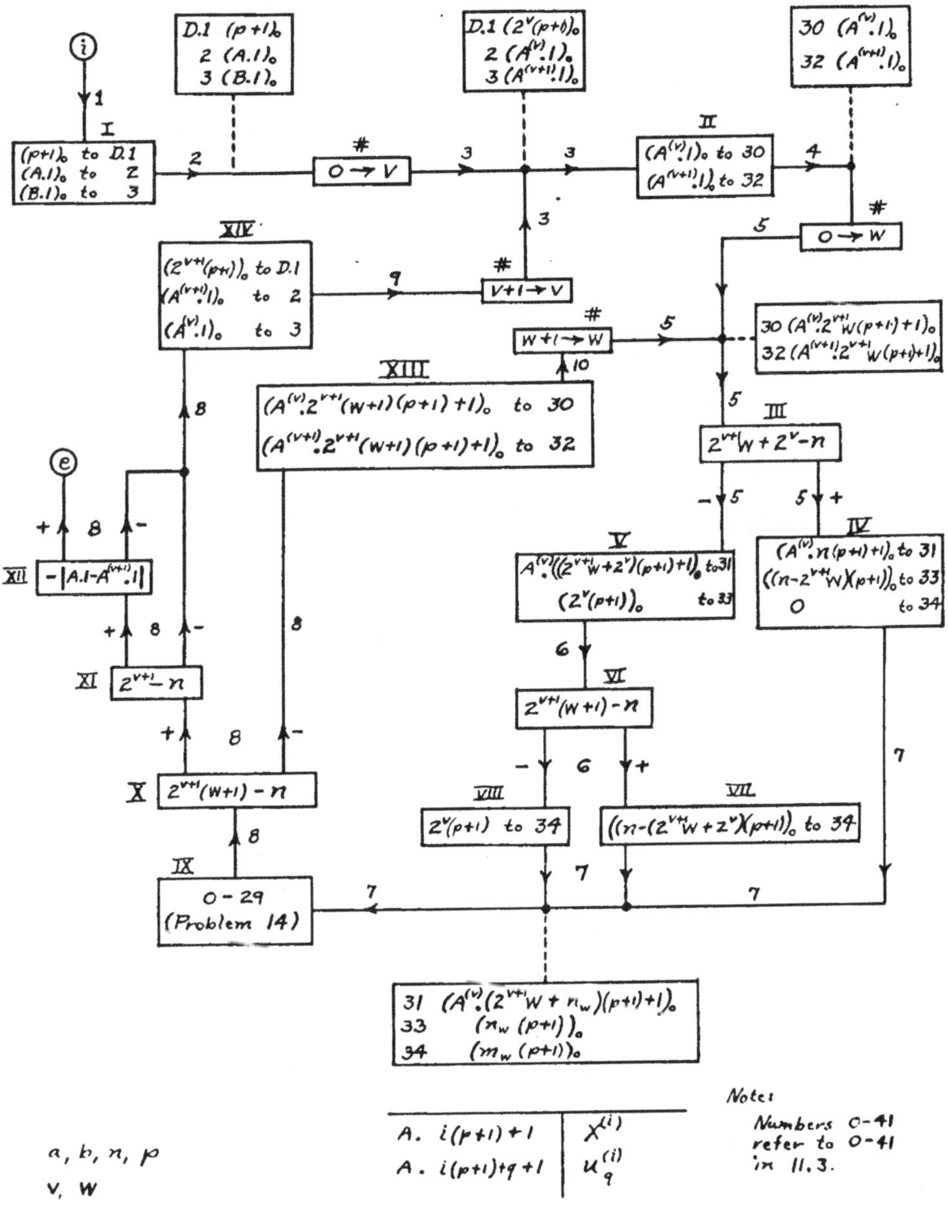

i
1
I
(p+1)0 to D.1
(A.1)0 to 2
(B.1)0 to 3
2
D.1 (p+1)0
2 (A.1)0
3 (B.1)0
#
0 → v
3
D.1 (2^v(p+1))0
2 (A^(v).1)0
3 (A^(v+1).1)0
3
II
(A^(v).1)0 to 30
(A^(v+1).1)0 to 32
4
30 (A^(v).1)0
32 (A^(v+1).1)0
#
0 → w
5
3
#
v+1 → v
9
XIV
(2^(v+1)(p+1))0 to D.1
(A^(v+1).1)0 to 2
(A^(v).1)0 to 3
#
w+1 → w
5
30 (A^(v).2^(v+1) w(p+1)+1)0
32 (A^(v+1).2^(v+1) w(p+1)+1)0
10
XIII
(A^(v).2^(v+1)(w+1)(p+1)+1)0 to 30
(A^(v+1).2^(v+1)(w+1)(p+1)+1)0 to 32
8
5
III
2^(v+1)w + 2^v − n
e
+ 8 −
XII
−|A.1−A^(v+1).1|
+ 8 −
XI
2^(v+1) − n
+ 8 −
X
2^(v+1)(w+1) − n
8
IX
0 − 29
(Problem 14)
8
− 5
5 +
V
A^(v).((2^(v+1)w+2^v)(p+1)+1)0 to 31
(2^v(p+1))0 to 33
IV
(A^(v).n(p+1)+1)0 to 31
((n−2^(v+1)w)(p+1))0 to 33
0 to 34
6
VI
2^(v+1)(w+1) − n
− 6 +
VIII
2^v(p+1) to 34
VII
((n−(2^(v+1)w+2^v)(p+1))0 to 34
7
7
7
7
31 (A^(v).(2^(v+1)w + n_w)(p+1)+1)0
33 (n_w (p+1))0
34 (m_w (p+1))0
A. i(p+1)+1
X^(i)
A. i(p+1)+q+1
u_q^(i)
a, b, n, p
v, w
Notes:
Numbers 0-41 refer to 0-41 in 11.3.

FIGURE 11.2

V,1	D.1		Ac	$(2^{\nu}(p+1))_0$
2	33	S	33	$(2^{\nu}(p+1))_0$
3	30		Ac	$(a^{(\nu)}+2^{\nu+1}w\ (p+1))_0$
4	D.1	h	Ac	$(a^{(\nu)}+(2^{\nu+1}w+2^{\nu})(p+1))_0$
5	31	S	31	$(a^{(\nu)}+(2^{\nu+1}w+2^{\nu})(p+1))_0$
(to VI,1)				
VI,1	D.2	h-	Ac	$((2^{\nu+1}w+2^{\nu})(p+1))_0$
2	D.1	h	Ac	$(2^{\nu+1}(w+1)(p+1))_0$
3	C.3	h-	Ac	$((2^{\nu+1}(w+1)-n)(p+1))_0$
4	VII,1	Cc		
(to VIII,1)				
VII,1	C.3		Ac	$(n(p+1))_0$
2	30	h-	Ac	$((n-2^{\nu+1}w)(p+1)-a^{(\nu)})_0$
3	D.2	h	Ac	$((n-2^{\nu+1}w)(p+1))_0$
4	D.1	h-	Ac	$((n-(2^{\nu+1}w+2^{\nu}))(p+1))_0$
5	34	S	34	$((n-(2^{\nu+1}w+2^{\nu}))(p+1))_0$
(to IX,1)				
VIII,1	D.1		Ac	$(2^{\nu}(p+1))_0$
2	34	S	34	$(2^{\nu}(p+1))_0$
(to IX,1)				
IX	(Problem 14: 0-29.)			
(to X,1)				
X,1	30		Ac	$(a^{(\nu)}+2^{\nu+1}w(p+1))_0$
2	D.2	h-	Ac	$(2^{\nu+1}w(p+1))_0$
3	D.1	h	Ac	$((2^{\nu+1}w+2^{\nu})(p+1))_0$
4	D.1	h	Ac	$(2^{\nu+1}(w+1)(p+1))_0$
5	C.3	h-	Ac	$((2^{\nu+1}(w+1)-n)(p+1))_0$
6	XI,1	Cc		
(to XIII,1)				
XI,1	D.1		Ac	$(2^{\nu}(p+1))_0$
2	D.1	h	Ac	$(2^{\nu+1}(p+1))_0$
3	C.3	h-	Ac	$((2^{\nu+1}-n)(p+1))_0$
4	XII,1	Cc		
(to XIV,1)				
XII,1	C.1		Ac	a_0
2	D.3	h-	Ac	$(a-a^{(\nu+1)})_0$
3	s.1	S	s.1	$(a-a^{(\nu+1)})_0$
4	s.1	-M	Ac	$-\lvert(a-a^{(\nu+1)})_0\rvert$
5	e	Cc		
(to XIV,1)				

XIII,1	30		Ac	$(a^{(v)}+2^{v+1}w\ (p+1))_0$
2	D.1	h	Ac	$(a^{(v)}+(2^{v+1}w+2^{v})(p+1))_0$
3	D.1	h	Ac	$(a^{(v)}+2^{v+1}(w+1)(p+1))_0$
4	30	S	30	$(a^{(v)}+2^{v+1}(w+1)(p+1))_0$
5	32		Ac	$(a^{(v+1)}+2^{v+1}w\ (p+1))_0$
6	D.1	h	Ac	$(a^{(v+1)}+(2^{v+1}w+2^{v})(p+1))_0$
7	D.1	h	Ac	$(a^{(v+1)}+2^{v+1}(w+1)(p+1))_0$
8	32	S	32	$(a^{(v+1)}+2^{v+1}(w+1)(p+1))_0$
(to III,1)				
XIV,1	D.1		Ac	$(2^{v}(p+1))_0$
2	D.1	h	Ac	$(2^{v+1}(p+1))_0$
3	D.1	S	Ac	$(2^{v+1}(p+1))_0$
4	D.2		Ac	$(a^{(v)})_0$
5	s.1	S	s.1	$(a^{(v)})_0$
6	D.3		Ac	$(a^{(v+1)})_0$
7	D.2	S	D.2	$(a^{(v+1)})_0$
8	s.1		Ac	$(a^{(v)})_0$
9	D.3	S	D.3	$(a^{(v)})_0$
(to II,1)				

The ordering of the boxes is I, II, III, V, VI, VIII, IX, X, XIII; XI, XIV; IV; VII; XII, and IX, IX, XIV, III, II must also be the immediate successors of IV, VII, XII, XIII, XIV, respectively. In addition, IX cannot be placed immediately after VIII, since IX,1 is 0 -- but IX must nevertheless be the immediate successor of VIII. All this necessitates the extra orders

IV,8	IX,1	C
VII,6	IX,1	C
VIII,3	IX,1	C
XII,6	XIV,1	C
XIII,9	III,1	C
XIV,10	II,1	C

Finally in order that X be the immediate successor of IX, the e of 0-41 (in 26) must be equal to X, 1.

We must now assign C.1-4, D.1-3, s.1 their actual values, pair the 76 orders I,1-6, II,1-4, III,1-5, IV,1-8, V,1-5, VI,1-4, VII,1-6, VIII,1-3, X,1-6, XI,1-4, XII,1-6, XIII,1-9, XIV,1-10 to 38 words, and then assign I,1 - XIV,10 their actual values. (IX is omitted, since it is contained in 0-41.) We wish to do this as a continuation of the code of 11.3. We will therefore begin with the number 42. Furthermore the contents of D.1-3, s.1 are irrelevant, like those of 37-41 there. We saw, in addition, that 30-35 (with the exceptions of 30,32) are also irrelevant. We must, however, make this reservation: 30-41 are needed while box IX operates, and during this period D.1-3 are relevantly occupied, but not s.1. s.1 is relevantly occupied only while boxes XII, XIV operate, and during this period 30-35 and 37-41 are irrelevant. Hence only s.1 can be made to coincide with one of these, say with 41. Summing all these things up, we obtain the following table:

I,1-6	42 -44'	VIII,1-3	54 -55	IV,1-8	70 -73'
II,1-4	45 -46'	X,1-6	55'-58	VII,1-6	74 -76'
III,1-5	47 -49	XIII,1-9	58'-62'	XII,1-6	77 -79'
V,1-5	49'-51'	XI,1-4	63 -64'	C.1-4	80 -83
VI,1-4	52 -53'	XIV,1-10	65 -69'	D.1-3	84 -86
				s.1	41

Now we obtain this coded sequence:

42	35 ,	84 S	57	84 h ,	82 h-	72	33 S ,	83
43	80 ,	85 S	58	63 Cc,	30	73	34 S ,	0 C
44	81 ,	86 S	59	84 h ,	84 h	74	82 ,	30 h-
45	85 ,	30 S	60	30 S ,	32	75	85 h ,	84 h-
46	86 ,	32 S	61	84 h ,	84 h	76	34 S ,	0 C
47	30 ,	85 h-	62	32 S ,	47 C	77	80 ,	86 h-
48	84 h ,	82 h-	63	84 ,	84 h	78	41 S ,	41 -M
49	70 Cc,	84	64	82 h-,	77 Cc	79	e Cc,	65 C
50	33 S ,	30	65	84 ,	84 h	80	a_0	
51	84 h ,	31 S	66	84 S ,	85	81	b_0	
52	85 h-,	84 h	67	41 S ,	86	82	$(n(p+1))_0$	
53	82 h-,	74 Cc	68	85 S ,	41	83	0	
54	84 ,	34 S	69	86 S ,	45 C	84	-	
55	0 C ,	30	70	82 ,	85 h	85	-	
56	85 h-,	84 h	71	31 S ,	30 h-	86	-	

For the sake of completeness, we restate that part of 0-41 of 11.3, which contains all changes and all substitutable constants of the problem. This is 26 and 30-41:

26	55 Cc', 10 Cc'	33	-	37	-
30	-	34		38	-
31		35	-	39	-
32		36	1_0	40	-
				41	-

The durations may be estimated as follows:

I: 225 μ, II: 150 μ, III: 200 μ, IV: 300 μ, V: 200 μ, VI: 150 μ, VII: 225 μ, VIII: 125 μ, X: 225 μ, XI: 150 μ, XII: 225 μ, XIII: 350 μ, XIV: 375 μ.

IX: The precise estimate at the end of 11.3 is

$$((575\,p + 1{,}125)\,(n_w + m_w) + 575)\ \mu.$$

Total: Put $\bar{v} = 2\left\langle {}^4\log n\right\rangle$, $\bar{\eta} = 2\left\langle {}^4\log n\right\rangle - \left\langle {}^2\log n\right\rangle = 0$ or 1, $\bar{w} = \left\langle \frac{n}{2^{\bar{v}+1}} \right\rangle$. Then the total is

$$I + \sum_{v=0}^{\bar{v}-1} \left\{ II + \sum_{w=0}^{\bar{w}-1} \left[\begin{array}{l} III + (IV^*) \text{ or } (V+VI+(VII^*) \text{ or } VIII))) \\ + IX + X + (XI^*) \text{ or } XIII) \end{array} \right] + XIV^{**} \right\} + XII \, (\bar{\eta}+1)$$

This is majorized by

$$(225 + \sum_{v=0}^{\bar{v}-1} \left\{ 150 + \sum_{w=0}^{\bar{w}-1} \left[\begin{array}{l} 200 + 200 + 150 + 125 + (575 \, p + 1,125) \times \\ \times (n_w + m_w) + 575 + 225 + 350 \end{array} \right] + 100 - 200 + 375 \right\} -$$
$$- 375 + 225 \, (\bar{\eta}+1)) \, \mu \, .$$

Since $\sum_{w=0}^{\bar{w}-1} (n_w + m_w) = n$, $\bar{\eta} \leq 1$, this is majorized by

$$(225 + \sum_{v=0}^{\bar{v}-1} \{(575 \, p + 1,125) \, n + 1,825 \langle \tfrac{n}{2^{v+1}} \rangle + 425\} + 75) \, \mu \, .$$

Since $\sum_{v=0}^{\bar{v}-1} \langle \tfrac{n}{2^{v+1}} \rangle \leq \sum_{v=0}^{\bar{v}-1} (\tfrac{n}{2^{v+1}} + 1) \leq n + \bar{v}$, this is majorized by

$$((575 \, p + 1,125) \, n \, \bar{v} + 1,825 \, n + 2,250 \, v + 300) \, \mu \, .$$

We can write this in this form:

$$((575 \, p + 1,200 \, (1 + \delta)) \, n \, \bar{v}) \, \mu \approx (.6 \, (p + 2 \, (1 + \delta)) \, n \, \bar{v}) \, m \, ,$$

where

$$\bar{v} = 2 \langle {}^4\!\log n \rangle \, ,$$
$$\delta = -.06 + \frac{1.52}{\bar{v}} + \frac{1.83}{n} + \frac{.25}{n \, \bar{v}} \, .$$

δ is a small and slowly changing quantity: For n = 100; 1,000; 10,000 (the last value is, of course, incompatible with the memory capacities which appear to be practical in the near future) we have $\bar{v}$ = 8; 10; 14 and hence δ = .15; .09; .05, respectively.

---

*) At most once among the w in $\sum_{w=0}^{\bar{w}-1}$. We disregard this effect for IV, which represents the shorter alternative. We replace VII, XI by their alternatives, VIII, XIII, respectively, inside the $\sum_{w=0}^{\bar{w}-1}$. This necessitates the corrections VII-VIII = 100 μ, XI-XIII = -200 μ, respectively, outside the $\sum_{w=0}^{\bar{w}-1}$. (Cf. the third and second terms from the right in the brackets { ... } of the next formula.)

**) Missing once among the v in $\sum_{v=0}^{\bar{v}-1}$. We may therefore subtract XIV = 375 μ as a correction outside the $\sum_{v=0}^{\bar{v}-1}$.

11.5 The meshing and sorting speeds of 11.3 and 11.4 are best expressed in terms of m per complex or of complexes per minute. The number of complexes in these two problems is n + m and n, respectively, hence the number of m per complex is $\approx .6\ (p+2)$ and $\approx .6\ (p+2(1+\delta)) \cdot 2 \langle {}^{4}\log n \rangle \approx 1.2\ (p+2.3) \langle {}^{4}\log n \rangle$, respectively. The number of complexes per minute obtains by dividing these numbers into 60,000, i.e. it is

$$(9) \qquad \frac{100,000}{p+2} \quad \text{and} \quad \frac{50,000}{(p+2.3)\langle {}^{4}\log n \rangle}, \text{ respectively.}$$

To get some frame of reference in which to evaluate these figures, one might compare them with the corresponding speeds on standard, electro-mechanical punch-card equipment.

Meshing can be effected with the I.B.M. collator, which has a speed of 225 cards per minute. Sorting can be effected by repeated runs through the I.B.M. (decimal) sorter. The sorting method which is then used is not based on iterated meshing, and hence differs essentially from our method in 11.4. Strictly construed, it requires as many runs through the sorter, as there are decimal digits in the principal number. This number of digits is usually between 3 and 8. Since we are dealing in our case with 40 binary digits, corresponding to 12 decimal digits, the use of the value 6 in such a comparison does not seem unfair. The sorter has a normal speed of 400 cards per minute, and it can unquestionably be accelerated beyond this level. 50% seems to be a reasonable estimate of this potential acceleration. Making the comparison on this basis, we have a speed of $\frac{400 \times 1.5}{6} = 100$ cards per minute. Hence we have these speeds for meshing and sorting in cards per minute:

$$(10) \qquad 225 \quad \text{and} \quad 100, \text{ respectively.}$$

A card corresponds to one of our complexes, since it moves as a unit in meshing and in sorting. Hence the speeds of (9) and (10) are directly comparable, and they give these ratios:

$$(11) \qquad \frac{(9)}{(10)} \approx \frac{444}{p+2} \quad \text{and} \quad \frac{500}{(p+2.3)\langle {}^{4}\log n \rangle}, \text{ respectively.}$$

A standard I.B.M. punch card has room for 80 decimal digits. This is equivalent to about 270 binary digits, i.e. somewhat less than 7 of our 40 binary digit numbers. It is, of course, in most cases not used to full capacity. Hence it is best compared to a complex with $p+1 \leqq 7$, i.e. $p = 1, \ldots, 6$. For n values from 100 to 1,000 seem realistic (in view of the probably available "inner" memory capacities, and assuming that no "outer" memory is used, cf. the second remark at the end of this section), hence $\langle {}^{4}\log n \rangle = 4, 5$. Consequently the ratios of (11) become

$$(11') \qquad \frac{(9)}{(10)} \approx 150 \text{ to } 55 \text{ and } 30 \text{ to } 15, \text{ respectively.}$$

In considering these figures the following facts should be kept in mind:

First: In using an electronic machine of the type under consideration for sorting problems, one is using it in a way which is least favorable for its specific characteristics, and most favorable for those of a mechanical punch card machine. Indeed, the coherence of a complex $X = (x; u_1, \ldots, u_p)$, i.e. the close connection between the movements of its principal number x and the subsidiary numbers $u_1, \ldots, u_p$, is guaranteed in a punch card machine by the physical identity and coherence of the punch card which it occupies, while in the electronic machine in question x and each $u_1, \ldots, u_p$ must be transferred separately; there is no intrinsic coherence between these items, and their ultimate, effective coherence results by synthesis from a multitude of appropriately coordinated individual transfers. This situation is very different from the one which exists in properly mathematical problems, where multiplication is a decisive operation, with the result that the extreme speed of the basic electronic organs can become directly effective, since the electronic multiplier is just as efficiently organized and highly paralleled as its mechanical and electromechanical counterparts. Thus in properly mathematical problems the ratio of speed is of the order of the ratio of the, say, .1 m multiplication time of an electronic multiplier and of the, say, 1 to 10 second multiplication time of an electromechanical multiplier, i.e., say, 10,000 to 100,000 -- while (11') above gave speed ratios of only 15 to 150.

Second: The "inner" memory capacities of the electronic machines that we envisage will hardly allow of values of n (and m) in excess of about 1,000. When this limit is exceeded, i.e. when really large scale sorting problems arise, then the "outer" memory (magnetic wire or tape, or the like) must also be used. The "inner " memory will then handle the problem in segments of several 100, or possibly up to 1,000, complexes each, and these are combined by iterated passages to and from the "outer" memory. This requires some additional coded instructions, to control the transfers to and from the "outer" memory, and slows the entire procedure somewhat, since the "outer" memory is very considerably less flexible and less fast available than the "inner" one. Nevertheless, this slowdown is not very bad: We saw in 11.3 that meshing requires .6 m per number. Magnetic wires or tapes can certainly be run at speeds of 20,000-40,000 (binary) digits per second, i.e. 500-1,000 (40 binary digit words or) numbers per second. This means 1-2 m per number. Thus the times required for each one of these two phases of the matter are of the same order of magnitude. In addition to this the "outer" memory is likely to be of a multiple, parallel channel type.

We will discuss this large scale sorting problem later, when we come to the use of the "outer" memory. It is clear, however, that it will render the comparison of speeds, (11'), somewhat less favorable.

Third: We have so far emphasized the unfavorable aspects of sorting with an electronic machine of the type under consideration-- i.e. one which is primarily an all-purpose, mathematical machine, not conceived particularly for sorting problems. Let us now point out the favorable aspects of the matter. These seem to be the main ones:

(a) All the disadvantages that we mentioned are relative ones (i.e. in comparison to the properly mathematical problems) -- electronic sorting should nevertheless be faster than mechanical, punch card sorting: In our above examples by factors of the order of 10 to 100.

(b) The results of electronic sorting appear in a form which is not exposed to the risks of the unavoidable human manipulations of punch cards: They are in the "inner" memory of the machine, which requires no human manipulation whatever; or in the "outer" memory, which is a connected, physically stable medium like magnetic wire or tape.

(c) The sorting operations can be combined with and alternated with properly mathematical operations. This can be done entirely by coded instructions under the inner control of the machine's own control organs, with no need for human intervention and no interruption of the fully automatic operation of the machine. This circumstance is likely to be of great importance in statistical problems. It represents a fundamental departure from the characteristics of existing sorting devices, which are very limited in their properly mathematical capabilities.

PLANNING AND CODING OF PROBLEMS
FOR AN
ELECTRONIC COMPUTING INSTRUMENT

By

Herman H. Goldstine   John von Neumann

Report on the Mathematical and Logical Aspects of an Electronic Computing Instrument

Part II, Volume III

The Institute for Advanced Study
Princeton, New Jersey
1948

PREFACE

This report was prepared in accordance with the terms of Contract No. W-36-034-ORD-7481 between the Research and Development Service, U. S. Army Ordnance Department, and The Institute for Advanced Study. It is another in a series of reports entitled, "Planning and Coding of Problems for an Electronic Computing Instrument", and it constitutes Volume III of Part II of that sequence. It is expected that Volume IV of this series will appear in the near future.

Herman H. Goldstine

John von Neumann

The Institute for Advanced Study
16 August 1948

TABLE OF CONTENTS

12.0 COMBINING ROUTINES

12.1 Each one of the problems that we have coded in the past Chapters 8-11 had the following properties: The problem was complete in the sense, that it led from certain unambiguously stated assumptions to a clearly defined result. It was incomplete, however, in another sense: It was certain in some cases and very likely in others, that the problem in question would in actual practice not occur by itself, as an isolated entity, but rather as one of the constituents of a larger and more complex problem. It is, of course, justified and even necessary from a didactical point of view, to treat such partial problems, problem fragments -- especially in the earlier stages of the instruction in the use of a code, or of coding per se. As the discussion advances, however, it becomes increasingly desirable to turn one's attention more and more from the fragments, the constituent parts, to the whole, In our present discussion, in particular, we have now reached a point where this change of emphasis is indicated, and we proceed therefore accordingly.

There are, in principle, two ways to effect this shift of emphasis from the parts to the whole.

The first way is to utilize the experience gained in the coding of simpler (partial) problems when one is coding more complicated (more complete) problems, but nevertheless to code all the parts of the complicated problem explicitly, even if equivalent simple problems have been coded before.

The second way is to code simple (partial) problems first, irrespective of the contexts (more complete problems) in which they may occur subsequently, and then to insert these coded sequences as wholes, when a complicated problem occurs of which they are parts.

We should illustrate both procedures with examples: This is not easy for the first one, because its use is so frequent that it is difficult to circumscribe its occurrences with any precision. Thus, if we had coded the calculation of the general third order polynomial, then any subsequent calculation involving (as a part) a third order polynomial, would offer such an example. Also, in view of Problem 1, any calculation involving a quotient of a second order and a first order polynomial would be an example. Problem 3, where the whole of Problem 1 is recoded as a part of the new coded sequence (but not Problem 2, where this is not done) is a specific instance.

Examples of the second procedure are more clearly identifiable. Problem 12 was used in this sense as a part of Problem 13.a and of Problem 13.b, and also, after some modifications, as a part of Problem 13.c. Problem 14 was used as a part of Problem 15. In addition it is fairly clear that all of the Problems 4-11 and 13-15 must be intended as parts of more complicated problems, and that it would be very convenient not to have to recode any one of them when it is to be used as part of another problem, but to be able to use it more or less unchanged, as a single entity.

12.2 The last remark defines the objective of this chapter: We wish to develop here methods that will permit us to use the coded sequence of a problem, when that problem occurs as part of a more complicated one, as a single entity, as a whole, and avoid the need for recoding it each time when it occurs as a part in a new context, i.e. in a new problem.

The importance of being able to do this is very great. It is likely to have a decisive influence on the ease and the efficiency with which a computing automat of the type that we contemplate will be operable. This possibility should, more than anything else, remove a bottleneck at the preparing, setting up, and coding of problems, which might otherwise be quite dangerous.

This principle must, of course, be applied with certain common sense limitations: There will be "problems" whose coded sequences are so simple and so short, that it is easier to recode them each time when they occur as parts of another problem, than to substitute them as single entities -- i.e. where the work of recoding the whole sequence is not significantly more than the work necessitated by the preparations and adjustments that are required when the sequence is substituted as a single entity. (These preparations and adjustments constitute one of the main topics of this chapter, cf. 12.3-12.5.) Thus the examples of the first procedure discussed in 12.1 above are instances of problems that are "simple" and "short" in this sense.

For problems of medium or higher complexity, however, the principle applies. It is not easy to name a precise lower limit for the complexity, say in terms of the number of words that make up the coded sequence of the problem in question. Indeed, this lower limit cannot fail to depend on the precise characteristics of the computing device under consideration, and quite particularly on the properties of its input organ. Also, it can hardly be viewed as a quite precisely defined quantity under any conditions. As far as we can tell at this moment, it is probably of the order of 15-20 words for a device of the type that we are contemplating (cf. the fourth remark in 12.11).

These things being understood, we may state that the possibility of substituting the coded sequence of a simple (partial) problem as a single entity, a whole, into that one of a more complicated (more complete) problem, is of basic importance for the ease and efficiency of running an automatic, high speed computing establishment in the way that seems reasonable to us. We are therefore going to investigate the methods by which this can be done.

12.3 We call the coded sequence of a problem a *routine*, and one which is formed with the purpose of possible substitution into other routines, a *subroutine*. As mentioned above, we envisage that a properly organized automatic, high speed establishment will include an extensive collection of such subroutines, of lengths ranging from about 15-20 words upwards. I.e. a "library" of records in the form of the external memory medium, presumably magnetic wire or tape. The character of the problems which can thus be disposed of in advance by means of such subroutines will vary over a very wide spectrum -- indeed a much wider one than is now generally appreciated. Some instances of this will appear in the subsequent Chapters 13 and 14. The discussions in those chapters will, in particular, give a more specific idea of what the possibilities are and what aims possess, in our opinion, the proper proportions.

Let us now see what the requirements and the difficulties of a general technique are, if this technique is to be adequate to effect the substitution of subroutines into routines in the typical situations.

12.4 The discussion of the precise way in which subroutines can be used, i.e. substituted into other routines, centers around the changes which have to be applied to a subroutine when it is used as a substituent.

These changes can be classified as follows:

Some characteristics of the subroutine change from one substitution of the subroutine (into a certain routine) to another one (into another routine), but they remain fixed throughout all uses of the subroutine within the same substitution (i.e. in connection with one, fixed routine). These are the changes of the *first kind*. Other characteristics of the subroutines may even vary in the course of the successive uses of the subroutine within the same substitution. These are the changes of the *second kind*.

Thus the order position at which a subroutine begins is constant throughout one substitution (i.e. routine, or, equivalently, one larger problem of which the subroutine's problem is part), but it may have to vary from one such substitution or problem to another. The first assertion is obviously in accord with what will be considered normal usage, the second assertion, however, needs some further elaboration.

If a given subroutine could only be used with its beginning at one particular position in the memory, which must be chosen in advance of all its applications, then its usefulness would be seriously limited. In particular, the use of several subroutines within one routine would be subject to very severe limitations. Indeed, two subroutines could only be used together, if the preassigned regions that they occupy in the memory do not intersect. In any extensive "library" of subroutines it would be impossible to observe this for all combinations of subroutines simultaneously. On the other hand, it will be hard to predict with what other subroutines it may be desirable to combine a given subroutine in some future problem. Furthermore, it will probably be very important to develop an extensive "library" of subroutines, and to be able to use it with great freedom. All solutions of this dilemma that are based on fixed positioning of subroutines are likely to be clumsy and of insufficient flexibility.

Hence we should postulate the variability of the initial order position of a subroutine from one substitution to another. Consequently this is an example of a change of the first kind. This requires corresponding adjustments of all references made in orders of the subroutine to definite (order or storage) positions within the subroutine, as they occur in the final form (the final enumeration) of its coding. These adjustments are, therefore, changes of the first kind.

The parameters or free variables of the problem that is represented by the subroutine (cf. 7.5) will, on the other hand, usually change from one use of the subroutine (within the same substitution, i.e. the same main routine or problem) to another. The same is true for the order position in the main routine, from which the control has to continue after the completion (of each particular use) of the subroutine. Since the subroutine sends the control after its completion to e (this is the notation that we have used in all our codings up to now, and we propose to continue using it in all subsequent codings), this observation can also be put as follows: The actual value of e will usually change from one use of the subroutine to another.

These remarks imply, that the parameters of the subroutines problem, as well as the actual value of its e, will usually undergo changes of the second kind.

12.5 All the changes that the use of a subroutine in a given substitution requires can be effected by the routine into which it is being substituted, i.e. by including appropriate coded instructions into that routine. For changes of the second kind this is the only possible way. For changes of the first kind, however, it is not necessary to put this additional load on the main routine. In this case the changes can be effected as preparatory steps, before the main routine itself is set in motion. Such preparations might be effected outside the machine (possibly by manual procedures, and possibly by more or less automatic, special, subsidiary equipment). It seems, however, much preferable to let the machine itself do it by means of an extra routine, which we call the *preparatory routine*. We will speak accordingly of an *internal preparation* of subroutines, in contradistinction to the first mentioned outside process, the *external preparation* of subroutines. We have no doubt that the internal preparation is preferable to the external one in all but the very simplest cases.

Thus changes of the first kind are to be effected by preparatory routines, which will be discussed further below. Changes of the second kind, as we have pointed out already, have to be effected by the main routine itself (into which the subroutine is being substituted): Before each use that the routine makes of the subroutine, it must appropriately substitute the quantities that undergo changes of the second kind (the parameters of the subroutines problem and the actual value of its e, cf. the discussion in 12.4), and then send the control to the beginning of the subroutine (usually by an unconditional transfer order). It may happen that some of these quantities remain unchanged throughout a sequence of successive uses of the subroutine. In this case the corresponding substitutions need, of course, be effected once, jointly for the entire sequence. If this sequence includes all uses of the subroutine within the routine, then the substitutions in question need only be performed once in the entire routine, at any sufficiently early point in it. In this last case we are, of course, really dealing with changes of the first kind, and the quantities in question could be dealt with outside the main routine, by a preparatory routine. It is, however, sometimes preferable to view this case as an extreme, degenerate form of a change of the second kind, or at any rate to treat it in that way.

This discussion should, for the time being, suffice to clarify the principles of the classification of subroutine changes, and of the effect which they (specifically: the changes of the second kind) have on the arrangements in the main routine (into which the subroutine is being substituted). We now pass to the discussion of the preparatory routine, which effects the essential changes of the first kind: The adjustments that are required in the subroutine by the variability of its initial order position.

12.6 Assume that a given subroutine has been coded under the assumption that it will begin at the order position a. (I.e. at the left-hand order of the word a. To simplify matters, we disregard the possibility that it may begin at the right-hand order of the word a. In our past codings we had usually a = 0, excepting Problem 2 where a = 100, Problems 13,a-c where a = 52, and Problem 15 where a = 42.) The orders that are contained in this subroutine can now be classified as follows:

First: The order contains no reference to a memory position x. It is then one of the orders 10, 20, 21 of Table II.

Second: The order contains a reference to a memory position x, but the place of this x is irrelevantly occupied in the actual code of the subroutine. In this case the subroutine itself must substitute appropriately for x, before the control gets to the order in question. I.e. some earlier part of the subroutine must form the substitution value for x, and substitute it into the order.

Third: The order contains a reference to a memory position x, the place of this x is relevantly occupied in the actual code of the subroutine, and this actual value of x corresponds to a memory position not in the subroutine.

Fourth: Same as the third case, except that the actual value of x corresponds to a memory position in the subroutine.

Fifth: One of the preceding cases, but at some point the subroutine treats the order or its x as if it were irrelevantly occupied, i.e. it substitutes there something else. ----

Assume next, that the subroutine, although coded as if it began at a, is actually to be used beginning at $\bar{a}$. This necessitates certain changes, which are just the ones that the preparatory routine has to effect, in the sense of the concluding remark of 12.5. Our above classification of the orders of the subroutine permits us to give now an exact listing of these changes.

Orders of the first and of the third kind require clearly no change. The same is true of the orders of the second kind if they produce x's which correspond to memory positions not in the subroutine. And even if x's are produced which correspond to memory positions in the subroutine, no change is necessary if the following rule has been observed in coding the subroutine: a was used explicitly in forming the x that corresponds to positions in the subroutine, and it was stored not as the actual quantity a, but as a parameter of the subroutine's problem. If it is then understood, that this parameter should have the value $\bar{a}$, then it will be adequately treated as a parameter in the sense of 12.5. Indeed, it represents that degenerate form of a change of the second kind, which can also be viewed as a change of the first kind, as discussed in 12.5. Thus it might be treated by a special step in the preparatory routine, but we prefer to assume, in order to simplify the present discussion, that it is handled as a parameter (of the subroutine) by the main routine. In this way the orders of the second kind require no change either (by the preparatory routine).

Orders of the fourth kind clearly require increasing their x by $\bar{a} - a$.

Orders of the fifth kind behave like a combination of an order of one of the four first kinds with an order of the second kind. Since all of these orders are covered by the measures that emerged from our discussion of the four first kinds of orders, it ensues that the orders of the fifth kind are automatically covered, too, by those measures.

Thus the preparatory routine has precisely one task: To add $\bar{a} - a$ to the x of every order of the fourth kind in the subroutine.

12.7 The next question is this: By what criteria can the preparatory routine recognize the orders of the fourth kind in the subroutine?

Let ℓ be the length of the subroutine. By this we mean the number of words, both orders and storage, that make it up. We include in this count all those words which have to be moved together when the final code (final enumeration) of the subroutine is moved (i.e. when its initial order position is moved from a to $\bar{a}$), and no others. The count is, of course, made on the final enumeration. In this sense a word counts fully, even if it contains a dummy order (e.g. 14 in Problem 6, and 6 in Problem 10 or 74, 91 in Problem 13.b). On the other hand storage positions which are being referred to, but which are supposed to be parts of some other routine, already in the machine (i.e. of the main routine, or of another subroutine) do not count (e.g. the storage area A in Problem 3 or the storage area A in Problem 10).

For an order of the fourth kind x must have one of the values $a, \ldots, a + \ell - 1$, i.e. it must fulfill the condition

(1) $$a \leqq x < a + \ell .$$

For an order of the third kind x will not fulfill this condition. For orders of the first and of the second kind the place of x is inessentially occupied. Concerning its relation to condition (1) we can make the two following remarks:

First: We can stipulate, that in all orders where the position of x is inessentially occupied, x should actually be put in with a value x^0 that violates (1). This is a perfectly possible convention. The simplest ways to carry it into effect are these:

Let x^0 always have the smallest value or always have the largest value that is compatible with its 12-digit character. (Regarding the latter, cf. section 6.2.) I.e. $x^0 = 0$ or $x^0 = L-1$, where L-1 is the largest 12-digit integer: $L = 2^{12} = 4{,}096$. Then all subroutines must fulfill $a \neq 0$ or $a + \ell \neq L$, respectively (in order that (1) be violated).

These rules are easy to observe. We chose a = 0 in most of our codes, hence we might prefer the second rule, but this is a quite unimportant preference.

Second: If an order of the first or of the second kind has an x which fulfills (1), and the order is thereupon mistakenly taken (by the preparatory routine) for one of the fourth kind, and its x is increased by $\bar{a} - a$, this need not matter either. Indeed: The place x is irrelevantly occupied, hence changes which take place there before the subroutine begins to operate do not matter. There is, however, one possible complication: Adding $\bar{a} - a$ to this (inessential) x may produce a carry beyond the 12 digits that are assigned to x. (Regarding these 12 digits cf. above, and also orders 18, 19 of Table II and the second remark among the Introductory Remarks to Chapter 10. The carry in question will occur if $\bar{a} - a > 0$ and $x \geq L - (\bar{a} - a)$, or if $\bar{a} - a < 0$ and $x < -(\bar{a} - a)$; $L = 2^{12}$, cf. above.) Such a carry affects the other digits of the order, and thereby modifies its meaning in an undesirable way. This complication can be averted by special measures that paralyze carries of the type in question, but we will not discuss this here. No precautions are needed, if we see to it that no such carries occur. (I.e. if we observe $-(\bar{a} - a) \leqq x \leqq L - (\bar{a} - a)$ for the inessential x, cf. above.) ----

In view of these observations we may accept (1) as the criterium defining the orders of the fourth kind. We will therefore proceed on this basis.

12.8 We have to derive the preparatory routines which are needed to make subroutines effective. For didactical reasons, we begin with a preparatory routine which can only be used in conjunction with a single (but arbitrary) subroutine. Having derived such a *single subroutine preparatory routine,* we can then pass to the more general case of a preparatory routine which can be used in conjunction with any number of subroutines. This is a *general,* or *multiple subroutine preparatory routine.* The point in all of this is, of course, that both kinds of preparatory routines need only be coded once and in advance -- they can then be used in conjunction with arbitrary subroutines.

We state now the problem of a single subroutine preparatory routine. This includes a description of the subroutine, in which we assume that the subroutine has been coded in conformity with the (not a priori necessary) conventions that we found convenient to observe in our codings in these reports. It does not seem necessary to discuss at this place possible deviations from these conventions, and the rather simple ways of dealing with them.

PROBLEM 16.

A subroutine Σ consisting of I consecutive words, of which the k first ones are (two) order words, is given. (Concerning the definition of the length of a subroutine, cf. the beginning of 12.7. The subroutine under consideration may also make use of stored quantities, or of available storage capacity, outside this sequence of I words -- or rather of the $I - k$ last ones among them. We need not pay any attention to such outside positions in this problem.) This subroutine is coded as if it began at the memory position a. Actually, however, it is stored in the memory, beginning at the position $\bar{a}$. It is desired to modify it so that its coding conform with its actual position in the memory. ----

Our task consists in scanning the words from $\bar{a}$ to $\bar{a}+k-1$, to inspect in each one of these words the two orders that it contains, to decide for each order whether its x fulfills the condition (1) of 12.7; and in that (and only in that) case increase this x by $\kappa = \bar{a} - a$.

Let the memory position u be occupied by the word w_u. w_u is then an aggregate of 40 binary digits:

$$w_u = \{ w_u (1), w_u (2), \ldots, w_u (40) \} .$$

The two orders of which it consists are the two 20 digit aggregates

$$\{w_u(1), \ldots, w_u(20)\} , \quad \{w_u(21), \ldots, w_u(40)\} ,$$

the two x in these orders are the two 12 digit aggregates

$$w_u^{I} = \{w_u(9), \ldots, w_u(20)\}, \; w_u^{II} = \{w_u(29), \ldots, w_u(40)\}$$

(cf. orders 18, 19 of Table II).

Reading w_u^{I}, w_u^{II} as binary numbers with the binary point at the extreme left (and an extra sign digit 0), the condition (1) of 12.7 becomes

$$(1) \qquad 2^{-12}a \leqq w_u^{I} < 2^{-12}(a + \lambda),$$

and

$$(2) \qquad 2^{-12}a \leqq w_u^{II} < 2^{-12}(a + \lambda).$$

Reading w_u as we ordinarily read binary aggregates, i.e. as a binary number with the binary point between the first and second digits from the left (the first digit being the sign digit) we can now express our task as follows: If (1) or (2) holds, we must increase w_u by $2^{-19}\kappa$ or $2^{-39}\kappa$, respectively. I.e.

$$(3) \qquad w_u' \begin{cases} = w_u + 2^{-19}\kappa & \text{if (1) holds,} \\ = w_u & \text{otherwise,} \end{cases}$$

$$(4) \qquad w_u'' \begin{cases} = w_u' + 2^{-39}\kappa & \text{if (2) holds,} \\ = w_u' & \text{otherwise.} \end{cases}$$

Note, that we may replace (2), for its use in (4), by

$$(2') \qquad 2^{-12}a \leqq w_u'^{II} < 2^{-12}(a + \lambda) ,$$

where

$$w'_u = \{ w'_u(1), w'_u(2), \ldots, w'_u(40)\} ,$$

$$w'^{II}_u = \{ w'_u(29), \ldots, w'_u(40)\} ,$$

since w_u and w'_u have by (3) the same digits with the positional values $2^{-20}, \ldots, 2^{-39}$, i.e. with the numbers 21, ..., 40.

The words w_u with which we have to deal are in the interval of memory locations $a + \kappa, \ldots, a + \kappa + k - 1$ (i.e. $\bar{a}, \ldots, \bar{a} + k - 1$, cf. above). Let this be the storage area O, we will index it with a $u = a + \kappa, \ldots, a + \kappa + k - 1$, so that O.u corresponds to u, and stores w_u. This u has the character of an induction index.

Further storage capacities are required as follows: u (as u_0) in A, the w_u under consideration (and w'_u after it) in B, the given data of the problem, a, ℓ, k, κ, in C. (It will be convenient to store them as $2^{-19}a$, $2^{-19}\ell$, $2^{-19}k$, $2^{-19}\kappa$. Regarding these quantities cf. also further below.) Storage will also have to be provided for various other fixed quantities (-1, 1_0, 2^{-7}, 2^{-12}, 2^{-32}), these too will be accomodated in C.

We can now draw the flow diagram, as shown in Figure 12.1. In coding it, we will encounter some deviations and complications which should be commented on.

a, ℓ, k, κ must occasionally be manipulated along the lines discussed in connection with the coding of Problem 13.a. Thus in the case of κ transitions to $2^{-39}\kappa$ and to κ_0 occur, and these would be rendered more difficult if we had to allow for the possibility $\kappa < 0$. In order to avoid this rather irrelevant complication, we assume

(5) $\qquad \kappa \geqq 0, \quad$ i.e. $\bar{a} \geqq a$.

This has the further consequence, that the difficulties referred to in 12.7 can be avoided by giving every irrelevant x the value 0 (because of the second remark in 12.7) or the value L - 1 (because of the first remark in 12.7). We also note this: (5) can be secured by putting all a = 0, i.e. by coding every subroutine as if it began at 0, but we will not insist here that this convention be made.

The conditions (1), (2') can be tested by testing the signs of the quantities

$$w^{I}_u - 2^{-12}a, \; w^{I}_u - 2^{-12}(a + \ell), \; w'^{II}_u - 2^{-12}a, \; w'^{II}_u - 2^{-12}(a + \ell)$$

or, equivalently, the signs of the quantities

$$2^{-7}w^{I}_u - 2^{-19}a, \; 2^{-7}w^{I}_u - 2^{-19}(a + \ell), \; 2^{-7}w'^{II}_u - 2^{-19}a, \; 2^{-7}w'^{II}_u - 2^{-19}(a + \ell) .$$

It is easily seen that replacing $2^{-19}a$, $2^{-19}(a+\mathcal{l})$ by a_0, $(a+\mathcal{l})_0$ vitiates these sign-criteria, but replacing

$$w_u^I = \{w_u(9), \ldots, w_u(20)\}$$

by

$$w_u^I + \varepsilon_u = \{w_u(9), \ldots, w_u(20), w_u(21), \ldots, w_u(40)\}$$

does not have this effect. It is convenient to use $w_u^I + \varepsilon_u$ in place of w_u^I.

Both quantities $w_u^I + \varepsilon_u$ and $w_u'^{II}$ must be read as binary numbers with the binary point at the extreme left, with an extra sign digit 0. Indicating this sign digit, too, we have

$$(6) \quad \begin{cases} w_u^I + \varepsilon_u = \{0, w_u(9), \ldots, w_u(40)\}, \\ w_u'^{II} = \{0, w_u'(29), \ldots, w_u'(40)\}. \end{cases}$$

With the same notations

$$(7) \quad \begin{cases} w_u = \{w_u(1), w_u(2), \ldots, w_u(40)\}, \\ w_u' = \{w_u(1), w_u(2), \ldots, w_u'(40)\}. \end{cases}$$

In order to get the $w_u^I + \varepsilon_u$, $w_u'^{II}$ of (6) from the w_u, w_u' of (7), it seems simplest to multiply w_u, w_u' by 2^{-32}, 2^{-12}, respectively, and to pick up $w_u^I + \varepsilon_u$ in the register (cf. order 11 of Table II). There is, however, one minor complication at this point: The register contains not the $w_u^I + \varepsilon_u$, $w_u'^{II}$ of (6), but the aggregates

$$(6') \quad \begin{cases} \{w_u(9), w_u(9), w_u(10), \ldots, w_u(40)\}, \\ \{w_u'(29), w_u'(29), w_u'(30), \ldots, w_u'(40)\}. \end{cases}$$

The simplest way to get from (6') to (6) is to sense the sign of each quantity of (6), and to add -1 to it if it proves to be negative (i.e. if the sign digit is 1). We will do this; it requires an additional conditional transfer in connection with each one of the two boxes III and VI. For this reason two boxes III.1 and VI.1, not shown in the flow diagram, will appear in our coding.

To conclude, it is convenient to change the position of IX somewhat (it follows upon VIII and absorbs part of it) and to absorb XI into X (it is replaced by X,9).

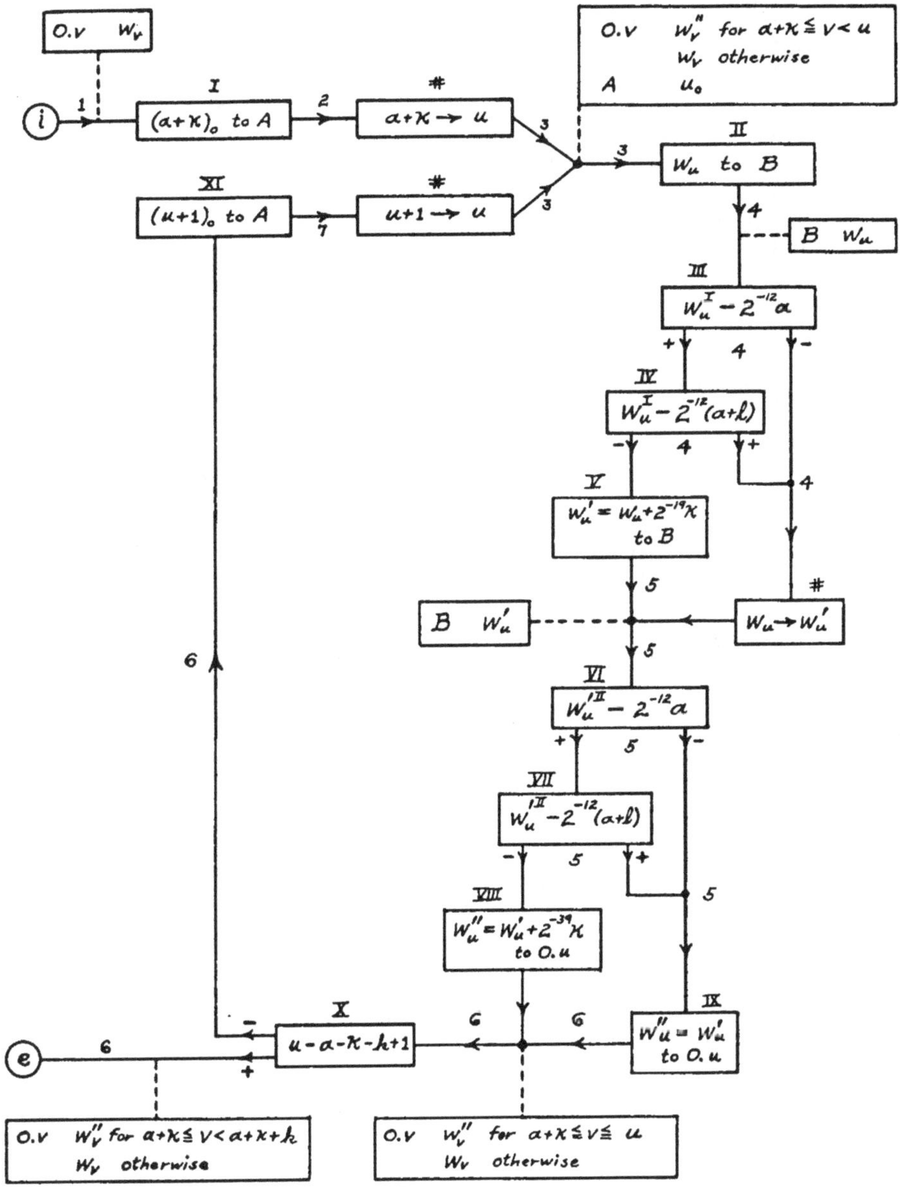

a, l, k, κ
u

FIGURE 12.1

The static coding of the boxes I-XI follows:

C.1	$2^{-19}a$			
C.2	$2^{-19}\kappa$			
I,1	C.1		Ac	$2^{-19}a$
2	C.2	h	Ac	$2^{-19}(a+\kappa)$
3	s.1	Sp'	s.1	$2^{-39}(a+\kappa)$
4	s.1	h	Ac	$(a+\kappa)_0$
5	A	S	A	$(a+\kappa)_0$
(to II,1)				
II,1	A		Ac	u_0
2	II,3	Sp	II,3	u
3	-			
[	u	]	Ac	w_u
4	B	S	B	w_u
(to III,1)				
C.3	2^{-32}			
III,1	C.3	R	R	2^{-32}
2	B	÷	R	$\{w_u(9), w_u(9), w_u(10), \ldots, w_u(40)\} =$ $= w_u(9) + w_u^I + \varepsilon_u$
3		A	Ac	$w_u(9) + w_u^I + \varepsilon_u$
4	III.1,1	Cc		
C.4	-1			
III,5	C.4	h	Ac	$w_u^I + \varepsilon_u$
(to III.1,1)				
C.5	2^{-7}			
III.1,1	C.5	R	R	2^{-7}
2	s.1	S	s.1	$w_u^I + \varepsilon_u$
3	s.1	x	Ac	$2^{-7}(w_u^I + \varepsilon_u)$
4	C.1	h-	Ac	$2^{-7}(w_u^I + \varepsilon_u) - 2^{-19}a$
5	IV,1	Cc		
(to VI,1)				
C.6	$2^{-19}\gamma$			
IV,1	C.6	h-	Ac	$2^{-7}(w_u^I + \varepsilon_u) - 2^{-19}(a+\gamma)$
2	VI,1	Cc		
(to V,1)				
V,1	C.2		Ac	$2^{-19}\kappa$
2	B	h	Ac	$w_u + 2^{-19}\kappa = w_u'$
3	B	S	B	w_u'
(to VI,1)				
C.7	2^{-12}			

VI,1	C.7	R	R	2^{-12}
2	B	v	R	$\{w_u'(29), w_u'(29), w_u'(30), \ldots, w_u'(40)\} =$ $= w_u'(29) + w_u'^{II}$
3		A	Ac	$w_u'(29) + w_u'^{II}$
4	VI.1,1	Cc		
5	C.4	h	Ac	$w_u'^{II}$
(to VI.1,1)				
VI,1.1	C.5	R	R	2^{-7}
2	s.1	S	s.1	$w_u'^{II}$
3	s.1	x	Ac	$2^{-7}\ w_u'^{II}$
4	C.1	h-	Ac	$2^{-7}\ w_u'^{II}\ -2^{-19}a$
5	VII,1	Cc		
(to IX,1)				
VII,1	C.6	h-	Ac	$2^{-7}\ w_u'^{II}\ -2^{-19}(a+\kappa)$
2	IX,1	Cc		
(to VIII,1)				
VIII,1	C.2		Ac	$2^{-19}\kappa$
2	s.1	Sp'	s.1	$2^{-39}\kappa$
3	s.1		Ac	$2^{-39}\kappa$
4	B	h	Ac	$w_u' + 2^{-39}\kappa = w_u''$
5	B	S	B	w_u''
(to IX,1)				
IX,1	A		Ac	u_0
2	IX,4	Sp	IX,4	u Sp
3	B		Ac	w_u''
4	-	S		
[	ι	S]	0.u	w_u''
(to X,1)				
X,1	C.1		Ac	$2^{-19}a$
2	C.2	h	Ac	$2^{-19}(a+\kappa)$
C.8	$2^{-19}k$			
X,3	C.8	h	Ac	$2^{-19}(a+\kappa+k)$
4	s.1	Sp'	s.1	$2^{-39}(a+\kappa+k)$
5	s.1	h	Ac	$(a+\kappa+k)_0$
6	s.1	S	s.1	$(a+\kappa+k)_0$
7	A		Ac	u_0
C.9	1_0			
X,8	C.9	h	Ac	$(u+1)_0$
9	A	S	A	$(u+1)_0$
10	s.1	h-	Ac	$(u-a-\kappa-k+1)_0$
11	e	Cc		
(to XI,1)				
XI,1	-			
(to II,1)				

Note, that the box XI required no coding, hence its immediate successor (II) must follow directly upon its immediate predecessor (X).

The ordering of the boxes is I, II, III, III.1, VI, VI.1, IX, X; IV, V; VII, VIII, and VI, IX, II must also be the immediate successors of V, VIII, X, respectively. This necessitates the extra orders

V,4	VI,1	C
VIII,6	IX,1	C
X,12	II,1	C

We must now assign A, B, C.1-9, s.1 their actual values, pair the 59 orders I,1-5, II,1-4, III,1-5, III.1,1-5, IV,1-2, V,1-4, VI,1-5, VI.1,1-5, VII,1-2, VIII,1-6, IX,1-4, X,1-12 to 30 words, and then assign I,1-X,12 their actual values. We wish to place this code at the end of the memory, so that it should interfere as little as possible with the memory space that is normally occupied by other subroutines and routines. Let us therefore consider the words in the memory backwards (beginning with the <u>last</u> word), and designate their numbers (in the reverse order referred to) by $\overline{1}$, $\overline{2}$, ... In this way we obtain the following table:

I,1-5	$\overline{42}$ - $\overline{40}$	VI.1,1-5	$\overline{30}$ - $\overline{28}$	VII,1-2	$\overline{17}'$ - $\overline{16}$
II,1-4	$\overline{40}'$ - $\overline{38}$	IX,1-4	$\overline{28}'$ - $\overline{26}$	VIII,1-6	$\overline{16}'$ - $\overline{13}$
III,1-5	$\overline{38}'$ - $\overline{36}'$	X,1-12	$\overline{26}'$ - $\overline{20}$	A	$\overline{12}$
III.1,1-5	$\overline{35}$ - $\overline{33}$	IV,1-2	$\overline{20}'$ - $\overline{19}$	B	$\overline{11}$
VI,1-5	$\overline{33}'$ - $\overline{31}'$	V,1-4	$\overline{19}'$ - $\overline{17}$	C,1-9	$\overline{10}$ - $\overline{2}$
				s.1	$\overline{1}$

Now we obtain this coded sequence:

$\overline{42}$	$\overline{10}$,	$\overline{9}$ h	$\overline{28}$	$\overline{17}$ Cc',	$\overline{12}$	$\overline{14}$	$\overline{11}$ h ,	$\overline{11}$ S
$\overline{41}$	$\overline{1}$ Sp',	$\overline{1}$ h	$\overline{27}$	$\overline{26}$ Sp,	$\overline{11}$	$\overline{13}$	$\overline{28}$ C',	- -
$\overline{40}$	$\overline{12}$ S ,	$\overline{12}$	$\overline{26}$	- S ,	$\overline{10}$	$\overline{12}$	-	
$\overline{39}$	$\overline{39}$ Sp',	-	$\overline{25}$	$\overline{9}$ h ,	$\overline{3}$ h	$\overline{11}$	-	
$\overline{38}$	$\overline{11}$ S ,	$\overline{8}$ R	$\overline{24}$	$\overline{1}$ Sp',	$\overline{1}$ h	$\overline{10}$	$2^{-19}a$	
$\overline{37}$	$\overline{11}$ x ,	A	$\overline{23}$	$\overline{1}$ S ,	$\overline{12}$	$\overline{9}$	$2^{-19}\kappa$	
$\overline{36}$	$\overline{35}$ Cc,	$\overline{7}$ h	$\overline{22}$	$\overline{2}$ h ,	$\overline{12}$ S	$\overline{8}$	2^{-32}	
$\overline{35}$	$\overline{6}$ R ,	$\overline{1}$ S	$\overline{21}$	$\overline{1}$ h-,	e Cc	$\overline{7}$	-1	
$\overline{34}$	$\overline{1}$ x ,	$\overline{10}$ h-	$\overline{20}$	$\overline{40}$ C',	$\overline{5}$ h-	$\overline{6}$	2^{-7}	
$\overline{33}$	$\overline{20}$ Cc',	$\overline{4}$ R	$\overline{19}$	$\overline{33}$ Cc',	$\overline{9}$	$\overline{5}$	$2^{-19}\gamma$	
$\overline{32}$	$\overline{11}$ x ,	A	$\overline{18}$	$\overline{11}$ h ,	$\overline{11}$ S	$\overline{4}$	2^{-12}	
$\overline{31}$	$\overline{30}$ Cc,	$\overline{7}$ h	$\overline{17}$	$\overline{33}$ C',	$\overline{5}$ h-	$\overline{3}$	$2^{-19}k$	
$\overline{30}$	$\overline{6}$ R ,	$\overline{1}$ S	$\overline{16}$	$\overline{28}$ Cc',	$\overline{9}$	$\overline{2}$	1_0	
$\overline{29}$	$\overline{1}$ x ,	$\overline{10}$ h-	$\overline{15}$	$\overline{1}$ Sp',	$\overline{1}$	$\overline{1}$	-	

The durations may be estimated as follows:

I: 200 μ, II: 150 μ, III: 250 μ, III.1: 270 μ, IV: 75 μ, V: 150 μ,
VI: 250 μ, VI.1: 270 μ, VII: 75 μ, VIII: 225 μ, IX: 150 μ, X: 450 μ.

$$\text{Total:}\quad I + \begin{bmatrix} II + III + III.1 + (\theta^{*)} \text{ or } IV \text{ or } IV + V) \\ + VI + VI.1 + (\theta^{*)} \text{ or } VII \text{ or } VII + VIII) \\ + IX + X \end{bmatrix} \times k =$$

$$\text{maximum} = \left(200 + \begin{bmatrix} 150 + 250 + 270 + 75 + 150 \\ + 250 + 270 + 75 + 225 \\ + 150 + 450 \end{bmatrix} k\right)\mu =$$

$$= (2{,}315\ k + 200)\ \mu \approx (2.3\ k + .2)\ m.$$

12.9 We now pass to the multiple subroutine preparatory routine. The requirements for such a routine allow several variants. We will consider only the basic and simplest one. It is actually quite adequate to take care of most situations involving the use of several subroutines -- even of very complicated ones. (Examples will occur in our future codings, in particular in Chapters 13 and 14.)

PROBLEM 17

Same as Problem 16 with this change: The modification is desired for I subroutines $\Sigma_1, \ldots, \Sigma_I$. The characteristic data for Σ_i (in the sense of Problem 16) are a_i, ℓ_i, k_i, κ_i. Each Σ_i is stored as its a_i, κ_i, and ℓ_i indicate (cf. Problem 16), the a_i, ℓ_i, k_i, κ_i $(i = 1, \ldots, I)$ are stored at 4I suitable, consecutive memory locations. ----

In order to be able to use the treatment of Problem 16 in 12.8, we assume in conformity with (5) in 12.8

(1) $\kappa_i \geqq 0$ for all $i = 1, \ldots, I.$

(Cf. also the other pertinent remarks loc. cit.)

i is the induction index, running from 1 to I. For each value of i we have to solve Problem 16. We can do this with the help of our coding of that problem, but we must substitute the data of the problem, a_i, ℓ_i, k_i, κ_i, into the appropriate places. Inspection of the coded sequence shows that these places are $\overline{10}$, $\overline{5}$, $\overline{3}$, $\overline{9}$, respectively.

We propose to place the coded sequence that we are going to develop immediately before that one of Problem 16, i.e. immediately before $\overline{42}, \ldots, \overline{1}$. Let P be the number of the memory location immediately before the coded sequence that we are going to develop -- i.e., if the length of that sequence is ℓ' and the total memory capacity is L' (both in terms of words), then

(2) $P = \overline{43 + \ell'} = L' - 43' - \ell'$

---

*) θ represents the possibility of going directly from III via 4, 5 to VI, or from VI via 5 to IX, respectively, with no other boxes intervening. We will use this same notation in similar situations in the future.

We will place the a_i, γ_i, k_i, κ_i ($i = 1, \ldots, I$), in the form $2^{-19}a_i$, $2^{-19}\gamma_i$, $2^{-19}k_i$, $2^{-19}\kappa_i$, immediately before these coded sequences, and in inverse order, i.e. at $P-4i+4$, $P-4i+3$, $P-4i+2$, $P-4i+1$, respectively.

The induction index i will be stored in the form $(P-4i+4)_0$ in the storage area A. The quantities P, I will be stored in the form P_0, $(P-4I)_0$ in the storage area B (1_0 will be needed and taken from $\bar{2}$).

We can now draw the flow diagram, as shown in Figure 12.2. The actual coding obtains from this quite directly, box V is absorbed into box II (it is replaced by II,10).

The static coding of the boxes I-V follows:

B.1	P_0			
I,1	B.1		Ac	P_0
2	A	S	A	P_0
(to II,1)				
II,1	A		Ac	$(P-4i+4)_0$
2	II,11	Sp	II,11	$P-4i+4$
3	$\bar{2}$	h-	Ac	$(P-4i+3)_0$
4	II,13	Sp	II,13	$P-4i+3$
5	$\bar{2}$	h-	Ac	$(P-4i+2)_0$
6	II,15	Sp	II,15	$P-4i+2$
7	$\bar{2}$	h-	Ac	$(P-4i+1)_0$
8	II,17	Sp	II,17	$P-4i+1$
9	$\bar{2}$	h-	Ac	$(P-4i)_0$
10	A	S	A	$(P-4i)_0$
11	-			
[	$P-4i+4$	]	Ac	$2^{-19}a_i$
12	$\overline{10}$	S	$\overline{10}$	$2^{-19}a_i$
13	-			
[	$P-4i+3$	]	Ac	$2^{-19}\gamma_i$
14	$\bar{5}$	S	$\bar{5}$	$2^{-19}\gamma_i$
15	-			
[	$P-4i+2$	]	Ac	$2^{-19}k_i$
16	$\bar{3}$	S	$\bar{3}$	$2^{-19}k_i$
17	-			
[	$P-4i+1$	]	Ac	$2^{-19}\kappa_i$
18	$\bar{9}$	S	$\bar{9}$	$2^{-19}\kappa_i$
(to III,1)				

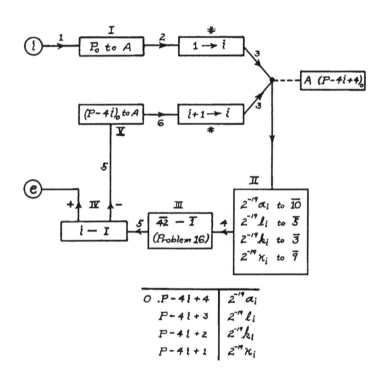

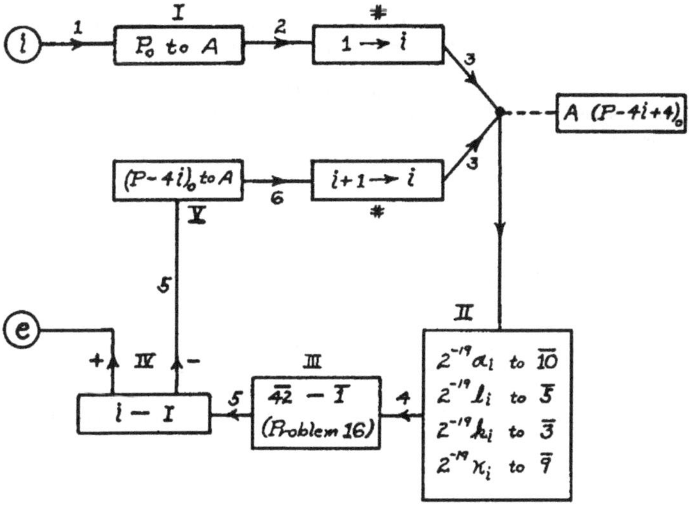

0 .P-4i+4	$2^{-19}a_i$
P-4i+3	$2^{-19}l_i$
P-4i+2	$2^{-19}k_i$
P-4i+1	$2^{-19}\kappa_i$

I
i

Note: Numbers $\overline{42}-\overline{1}$ refer to $\overline{42}-\overline{1}$ in 12.7

FIGURE 12.2

III (Problem 16: $\overline{42} - \overline{1}$.)

(to IV,1)

B.2	$(P-4I)_0$			
IV,1	B.2		Ac	$(P-4I)_0$
2	A	h-	Ac	$(4(i-I))_0$
3	c	Cc		

. (to V,1)

V	-

(to II,1)

Note, that box V required no coding, hence its immediate successor (II) must follow directly upon its immediate predecessor (IV).

The ordering of the boxes is I, II, III, IV, and II must also be the immediate successor of IV. In addition, III cannot be placed immediately after II, since III,1 is 42 but III must nevertheless be the immediate successor of II. All this necessitates the extra orders

II,19	III,1	C
IV,4	II,1	C

Finally, in order that IV be the immediate successor of III, the e of $\overline{42} - \overline{1}$ (in 21) must be equal to IV,1.

We must now assign A, B.1-2 their actual values, pair the 25 orders I,1-2, II,1-19, IV,1-4 to 13 words, and then assign I,1-IV,4 their actual values. (III is omitted, since it is contained in $\overline{42} - \overline{1}$.) We wish to do this as a (backward) continuation of the code of 12.8. In this way we obtain the following table:

I,1-2	$\overline{58} - \overline{58}'$	II,1-19	$\overline{57} - \overline{48}$	A	$\overline{45}$
		IV,1-4	$\overline{48}' - \overline{46}$	B.1-2	$\overline{44} - \overline{43}$

Thus, in terms of equation (2), $\mathcal{L}' = 16$ and

(2') $$P = \overline{59} = L' - 59 \; .$$

Now we obtain this coded sequence:

$\overline{58}$	$\overline{44}$, $\overline{45}$ S	$\overline{53}$	$\overline{2}$ h-, $\overline{45}$ S	$\overline{48}$	$\overline{42}$ C , $\overline{43}$
$\overline{57}$	$\overline{45}$, $\overline{52}$ Sp	$\overline{52}$	- , $\overline{10}$ S	$\overline{47}$	$\overline{45}$ h-, e Cc
$\overline{56}$	$\overline{2}$ h-, $\overline{51}$ Sp	$\overline{51}$	- , $\overline{5}$ S	$\overline{46}$	$\overline{57}$ C , - -
$\overline{55}$	$\overline{2}$ h-, $\overline{50}$ Sp	$\overline{50}$	- , $\overline{3}$ S	$\overline{45}$	-----
$\overline{54}$	$\overline{2}$ h-, $\overline{49}$ Sp	$\overline{49}$	- , $\overline{9}$ S	$\overline{44}$	P_0
				$\overline{43}$	$(P-4I)_0$

In addition $\overline{21}$ in $\overline{42}$ - $\overline{1}$ of 12.8 must read

$\overline{21}$ $\overline{1}$ h-, $\overline{48}$ Cc'

The durations may be estimated as follows:

I: 75 μ, II: 725 μ, IV: 150 μ.
III: The precise estimate at the end of 12.7 is

$$\text{maximum} = (2{,}315\ k_i + 200)\ \mu.$$

Total: $\text{I} + \sum_{i=1}^{I} (\text{II} + \text{III} + \text{IV}) =$

$$\text{maximum} = (75 + \sum_{i=1}^{I} (725 + 2{,}315\ k_i + 200 + 150))\ \mu =$$

$$= (2{,}315 \sum_{i=1}^{I} k_i + 1{,}075\ I + 75)\ \mu \approx$$

$$\approx (2.3 \sum_{i=1}^{I} k_i + 1.1\ I)\ m\ .$$

$\sum_{i=1}^{I} k_i$ is the total length of all subroutines. Hence it is necessarily $\leq L' \leq L = 2^{12} = 4{,}096$. Actually it is unlikely to exceed, even in very complicated problems, the order $\approx \frac{1}{4} L \approx 1{,}000$. 1.1 I is negligible compared to $2.3 \sum_{i=1}^{I} k_i$.
Hence

$$2{,}300\ m = 2.3 \text{ seconds}$$

is a high estimate for the duration of this routine.

12.10 Having derived two typical coded sequences for preparatory routines, it is appropriate to say a few words as to how their actual use can be contemplated. We do not propose to present a discussion of this subject to any degree of completeness -- we only wish to point out some of the most essential aspects. Since the routine of Problem 17 is more general than that one of Problem 16, we will base our discussion on the former.

The discussion of the use of preparatory subroutines is necessarily a discussion of a certain use of the input organs of the machine. We have refrained in our reports, so far, from making very detailed and specific assumptions regarding the characteristics of the input (as well as of the output) organs. At the present juncture, however, certain assumptions regarding these organs are necessary. On the basis of engineering developments up to date, such assumptions are possible on a realistic basis. Those that we are going to formulate represent a very conservative, minimum program.

Our assumptions are these:

As indicated in section 4.5, we incline towards the use of magnetic wire (soundtrack) as input (and output) medium. We expect to use it at pulse rates of about 25,000 pulses (i.e. binary digits) per second. We will certainly use several input (and output) channels, but for the purposes of the present discussion we assume a single one.

We assume that the contents of a wire can be "fed" into the machine, i.e. transferred into its inner, selectron memory, under manual control. We assume that we can then describe by manual settings

1) to which memory position the first word from the wire should go (the subsequent words on the wire should then go in linear order into the successive, following memory positions),

2) how many words from the wire should be so "fed".*)

We assume finally, that single words can also be "fed" directly into the machine by typing them with an appropriate "typewriter". (This "typewriter" will produce electrical pulses, and will be nearly the same as the one used to "write" on the magnetic tape.) We assume that we can determine the memory position to which the typed word goes by manual settings.

The last assumption, i.e. the possibility of typing directly into the memory, is not absolutely necessary. It is alternative to the previously mentioned "feeding" of words from a magnetic wire. When longer sequences of words have to be fed, the wire is preferable to direct typing. When single words, irregularly distributed, have to be fed, however, then feeding from appropriate wires would still be feasible, but definitely more awkward than direct typing. In addition, the possibility of direct typing into the memory is probably very desirable in connection with testing procedures for the machine. We are therefore assuming its availability.

These things being understood, we can describe the procedure of placing a, presumably composite, routine into the machine. It consists of the following steps:

First: There are one or more constituent routines: The main routine and the subroutines, where it is perfectly possible that the subroutines bear further subordination relationships to each other, i.e. are given as a hierarchy. All of these are coded and stored on separate pieces of magnetic wire.

These are successively fed into the machine, i.e. into the inner memory. The desired positions in the memory are defined by manual settings.

---

*) These operations should also be feasible under the "inner", electronic control of the machine. We will discuss this aspect of the input-output organ, and the logical, code orders which circumscribe it, in a subsequent paper (Part III in this series).

Second: The multiple subroutine preparatory routine (Problem 17) is also coded and stored on an individual magnetic wire. (It is, of course, assumed to be part of the library of wires mentioned in section 4.6.)

This is also fed into the machine, like the "constituent" routines of the first operation, described above. As observed in 12.8 and 12.9 it has to be positioned at the end of the memory.

Third: The constants of the multiple subroutine preparatory routine, i.e. of Problem 17, are typed directly into the memory. They are the following ones:

1) The number of subroutines, I, which is put in the form $(P-4I)_0'$ into the position

$$\overline{43} = L' - 43 = P + 16.$$

2) The 4I data which characterize the I subroutines: a_i, l_i, k_i, κ_i $(i = 1,\ldots,I)$, which are put in the form $2^{-19}a_i$, $2^{-19}l_i$, $2^{-19}k_i$, $2^{-19}\kappa_i$ into the positions P-4i+4, P-4i+3, P-4i+2, P-4i+1, respectively.

Fourth: The machine is set going, with the control set manually at the beginning of the preparatory routine ($\overline{58} = L'-48 = P+1$), and the adjustment of all subroutines to their actual positions (in the sense of 12.5-12.7) is thus effected.

Fifth: Any further adjustments which are necessitated by the relationships of the subroutines to each other (cf. the first operation, as described above, and the fifth remark in 12.11) are made by typing directly into the memory. ----

After all these operations have been carried out, the machine is ready to be set going on the (composite) routine of the problem itself The preparatory routine (in the 58 last memory positions) and its data (in the 4I preceding positions) are now no longer needed. I.e. these positions may now be viewed, from the point of view of the problem itself, as irrelevantly occupied. They are accordingly available for use in this sense.

12.11 We can now draw some conclusions concerning the setting-up procedure for a machine of the type contemplated, on the basis of the discussion of 12.10. We state these in the form of five successive remarks.

First: The pure machine time required to feed the (main and subsidiary) routines of the problem into the machine may be estimated as follows: A word consists of 40 pulses. For checking and marking purposes it will probably have to contain some additional pulses. With the systems that we are envisaging, a total of, say, 60 pulses per word will not be exceeded. With the speed of 25,000 pulses per second, as assumed in 12.10, this gives 2.4 m per word.

Thus the pure machine time of the first and second operations of 12.10 is 2.4 m per word.

The pure machine time of the third operation of 12.10 is, as we saw at the end of 12.9, 2.3 m per word.

We also observed at the end of 12.9, that even in very complicated problems the number of words thus involved is not likely to exceed 1000. This puts on the total time requirement on these counts an upper limit of 2.4 + 2.3 = 4.7 seconds, i.e. of less than 5 seconds.

Second: These time requirements are obviously negligible compared to the time consumed by the attendant manual operations: The placing of the magnetic wires into the machine, the setting of the (memory) position definitions, etc.

It follows therefore, that there is no need and no justification for any special routine-preparing equipment (other than the typing devices already discussed) to complement a machine of the type that we contemplate.

Third: Assuming a composite routine made up of ten parts, i.e. of a routine and nine subroutines, we have I = 9. This represents already a very high level of complication. The preparatory routine requires 58 + 4I words, i.e. for I = 9 94 words. This represents 2.3% of the total (selectron) memory capacity, if we assume that the latter is $L' = 2^{\cdot 12} = 4{,}096$.

Fourth: Each subroutine requires the direct, manual typing of four words into the machine (for a_i, λ_i, k_i, κ_i), as well as one for all subroutines together (for I). In addition the changes of the second kind in the sense of 12.4, i.e. those which the main routine must effect on the subroutine, require several words. Indeed, the sending of the control to the subroutine requires one order, i.e. half a word. Any number is sent there at the price of two orders (bringing the number to be substituted into the subroutine into the accumulator, substituting it into the subroutine) and of possibly one storage word (for the number to be substituted), i.e. of a total of one or two words. There will usually be three or more such number substitutions (the e of the subroutine, i.e. the memory position in the main routine from where the control is to continue after the completion of the subroutine, and two or more data for the subroutine). Thus five words for these changes is a conservative estimate.

A subroutine consumes therefore ten words in extra instructions, by a conservative estimate. It seems therefore, that the storage of a subroutine in a library of wires, in the sense of section 4.6, and its corresponding treatment as an individual entity becomes justified when its length in words is significantly larger than 10. A minimum length of 15-20 words would therefore seem reasonable.

To conclude this discussion, we observe that in making these estimates, we disregarded all operations other than the actual, manual typing of words (on wire or into the machine). This is legitimate, because the time and the memory requirements of the automatic operations that are involved are negligible, as we saw in our three first remarks.

Fifth: We pointed out in our description of the first operation in 12.10, that the various subroutines used in connection with a main routine, may bear further subordination relationships to each other. In this case they will also contain actual references to each other, and these will have to be adjusted to the actual positions of the subroutines in question in the memory. These adjustments may be made as changes of the second kind, in the sense of 12.4, by the routines involved. They may also be handled by special preparatory routines. We expect, however, that it will be simplest in most cases to take care of them by direct typing into the memory, as indicated in the description of the fifth operation in 12.10.

The adjusting of the references of a subroutine to itself, to the actual position of the subroutine in the memory, might also have been made by direct typing into the machine. We chose to do it automatically, however, by means of a preparatory routine, because these references are very frequent: The great majority of all orders in a subroutine contain references to this same subroutine. References to another subroutine, on the other hand, are likely to be rare and irregularly distributed. They are therefore less well suited to automatic treatment, by a special preparatory routine, than to ad hoc, manual treatment, by direct typing into the machine.

Actual examples of such situations in which it will also be seen that the proportions of the various factors involved are of the nature that we have anticipated here, will occur in the subsequent chapters, and in particular in Chapters 13 and 14.

ND - #0188 - 021224 - C0 - 229/152/10 - PB - 9781332178841 - Gloss Lamination